Jacky Yu與你分享醃漬釀酒的樂趣！

*Jacky Yu shares with you
the joy of pickling and wine-making.*

慢下來，
感受生活

我喜歡飲食，因為它是一種生活，一種態度，還蘊含着無盡的文化及無邊的領域，在烹飪的世界裏面，不單讓我可以盡情發揮，更讓我感受到無限的生活樂趣及分享的喜悅。

這是我第五本個人食譜，在主題上我很希望有所突破，除了研創菜式及享受美食之外，醃漬泡菜、浸釀果酒、弄果醬、調製醬汁等，這些都是我平日非常享受及喜歡做的事情。在這個生活節奏過分忙碌的大都市，偶爾慢下來，享受一下這種自得其樂的料理樂趣，實在是一種舒緩壓力，感受生活的好方法。

在這本書，我將多年來在這方面的食譜及經驗，在這裏徹底地與大家分享！

在烹飪的世界裏，我找到了自己的天地，以及存在價值，因為在這裏它有着無盡的元素與可能！

我家快樂農夫~
父親

　　人，如何可以活得快樂？

　　快樂，可以是一件相當高深的事！這
要看你是不是一個易於滿足的人，快樂其
實亦可以是非常簡單，而且它就在你的附
近，看你願不願意跟它靠近而已……

　　父親出生在排球之鄉台山，年輕時曾
當過國家隊排球運動員、體育教師，來到
香港之後竟然做了建築工人，在金門建築
公司當地盤判頭，一當便是三十多年直到
退休。

　　其實，父親天生是一個農夫，可能
自小在一個務農鄉鎮長大，他喜歡有關務
農耕種的所有事情，對園藝種植之類活動
非常熱愛，有空便喜歡到花墟流連，去西
貢的花場逛逛，以及喜歡研究很多農作物
種植的種種方法及常識。記得我們小時候
住的房子沒有露台，但家裏幾乎所有可以
擺放盆栽的地方都種滿了不同的植物，他
還可以在每個施工的地盤設法找出一塊可
以種植的空地，依照不同環境及季節種植
不同的蔬菜，有收成的時候便分給其他工
友。

十多年前，父親退休了。他一早已經計劃好一定要住在新界及可以種東西的地方，他買了「丁屋」，開始了退休後的農夫生活，他非常非常享受及滿足於這種生活，這也是他多年來的夢想及他一直所追求的。原本他只是在家裏的露台及天台種些自己喜歡的植物及蔬菜，但當村長見到他所種出來的「農作物」如此有水準及熱愛，便索性劃了一塊丁地給他種植。現在，在父親的那塊農地裏，種植了很多各式各樣的農作物，我家一年四季的蔬菜瓜果基本上都是自給自足，我每次回家吃飯，父親都會到菜田隨手摘一些蔬菜回家，然後馬上烹調，味道既新鮮、感覺又溫暖。

無可否認，父親對耕種的確是很有天分的，甚麼植物去到他手上都可以茁壯成長，而且他還有很多花樣，他曾經將檸檬樹與柚子樹接枝，結果長出了柚子形狀及大小但是檸檬味道的「柚子檸檬」。父親亦很有環保意識，他從不使用化學肥料，他會將吃剩的爛菜葉、果皮及收集回來的枯葉之類東西儲存起來發酵，變成有機肥料，種出來的農作物保證安全及健康！村裏的左鄰右里都很喜歡他，因為父親是一個很老實、很喜歡與人分享的人，種植出來的蔬菜及其他農作物，他大部分都是分給鄰居。而且父親為人樂觀隨和，凡事都很看得開，以前家裏發生過的很多大小困難，他都能坦然面對。

每個人對快樂標準不一，對我父親來說：快樂就是一家人平安健康，能夠過一個滿足到自己的簡單生活。他就覺得已經非常的快樂！

在此，祝爸爸身體永遠健康！快樂！

造好的食物，
應如何保存？

烹飪文化越來越普及，很多人都喜歡在家DIY，既是生活樂趣又可以吃得更安心。一般最常在家裏製作的果醬或是中式醬料。如果保存得不妥當，就會很易滋生細菌，影響健康。如食物若是在一星期左右吃完的話，裝入洗淨、乾燥並有蓋子的容器，蓋好放入雪櫃即可。但若是大量製造，想送給朋友，保存期需要2-3個月或更長的話，你便一定要經過高溫消毒、殺菌或做抽真空處理了。其實做法也很簡單，在家裏亦很容易做到！以下的兩種方法是我經常使用而效果亦相當衛生、安全！

先將容器消毒：

容器清洗乾淨後，煲滾一鍋水，將容器連蓋放入鍋內(水要能蓋過容器)，用沸水焓約5-8分鐘，小心用鉗子夾起，瀝乾水分，然後放在乾淨的毛巾上，讓它自然涼卻及吹乾水分。

方法1：將容器倒轉的保存方法

將煮好的醬料趁熱倒入已消毒的容器內，裝滿至距離容器頂部約2公分的容量，預留些許空氣可以流通的空間，輕輕蓋上蓋子，然後馬上將容器倒轉置放，直至容器自然冷卻。這個方法比較簡單，原理及效果與下面介紹的方法2相同，但必須將醬料煮好後第一時間、溫度仍在攝氏80度或以上時放入容器，因為食物在攝氏4度至65度之間時稱之為危險區域，食物在這溫度之間容易滋生細菌及容易變壞。

方法2：真空處理法

　　將煮好的醬料趁熱倒入已消毒的容器內，裝滿至距離容器頂部約2公分的容量，預留些許空氣可以流通的空間，輕輕蓋上蓋子。取一平底鍋，放入容器，鍋內注入清水，水要注到容器約三分之二的位置，將水煲滾後以慢火將容器焓30分鐘，用鉗子將容器夾起，瀝乾水分，讓它自然冷卻後便可毋須放入雪櫃，在常溫狀態下一般的醬料都至少可以保存6個月至1年的時間(要視乎醬料成分及濃度)。

　　這種真空處理的原理是，在焓煮容器的過程中，除了可讓高溫將食物消毒之外，還令容器內的空氣膨脹，當容器冷卻之後，容器內的空氣便會收縮，更會將瓶蓋吸緊至凹陷的效果，形成與外間空氣絕緣，容器內真空的狀態，這樣食物便可以得到長時間的保存了。

貼士 🐝 ━━━━━━━━━━━━━━━━━━━━━━

1. 食物開蓋後便會與外間空氣接觸，容易滋生細菌，因此開蓋後的食物便應放入雪櫃保存及盡快吃完。

2. 當容器裝滿食物蓋上瓶蓋後，千萬不要貪快將它放入冷水中或淋凍水以令它加速冷卻，因為這樣很易令容器爆裂，造成危險。

3. 完成真空處理後的容器應貼上標籤，注明製造日期，以便日後易於識別保存期限。

4. 瓶蓋不要擰得太緊，應留少許空間，以免熱脹冷縮後瓶蓋會吸得太緊難以擰開。

5. 市面上都可以買到一些專門可以做真空處理的玻璃瓶子，瓶蓋已預留可以凹陷的設計及識別，相當好用，效果亦非常理想，而且可以循環再用，非常環保！

目
錄

summer

果酒

番石榴酒......62

私房楊梅酒......64

楊梅桂花陳酒......66

日本水蜜桃酒......68

乾果酒......70

黑桑椹霖酒......74

私房紅莓酒......76

巨峰提子酒......78

日本王林蘋果酒......80

迷人櫻桃酒......84

私房李子酒......86

黑桑棗霖酒......88

荔枝桂花陳酒......90

果醬

熱情果菠蘿果醬......92

玫瑰水蜜桃果醬......95

蜂蜜奇異果果醬......98

雲呢嗱芒果果醬......100

黑糖楊桃果醬......102

桂花陳酒荔枝果醬......104

尊貴巨峰提子果醬......106

話梅菠蘿果醬......108

仁稔

被遺忘的滋味__阿姨仁稔......110

仁稔汁涼拌皮蛋豆腐......114

泡菜

星洲娘惹泡菜......116

青木瓜

傳統糖醋醃青木瓜......120

熱情果醃青木瓜......122

spring

柑桔

柑桔食療......16

過年後齊齊醃鹹柑桔......17

醃鹹柑桔......18

鹹柑桔麵豉醬蒸烏頭......20

金銀柑桔果醬......22

私房柑桔酒......24

桂花柑桔酒......25

梅子

浸泡私房梅子酒......27

香濃黑糖梅酒......30

梅酒凍......32

呷醋有益......35

私房梅子醋......36

梅子蜂蜜......38

私房黑糖梅子漿......40

私房鹹水梅......44

梅子蒸花蟹......46

子薑梅子燜雞......48

草莓

草莓果醬......50

果醬乳酪......53

草莓蜂蜜......54

火紅催情草莓果酒......56

子薑
三色醃子薑......124

冬瓜
麵豉冬瓜醬......129
麵豉冬瓜醬蒸鯧魚......132
麵豉冬瓜醬蒸肉餅......134

青芒
泰式醃青芒......136
泰式醃青芒炒牛肉......138
泰式青芒果海鮮沙律......140

白玉涼瓜
沖繩黑糖醃白玉涼瓜......142

autumn

蘋果
自製蘋果果醋......146
蘋果醋醃車喱茄......149
車喱茄果醋啫喱......152
車喱茄果醋特飲......154

番茄
低溫松露油風乾番茄......155
冰鎮薄荷梅子番茄......158

鹹蛋
私房五香鹹蛋......160
五香鹹蛋炒涼瓜......164

三文魚
清酒香草醬油漬三文魚......166
胡椒海鹽醃三文魚......170

胡椒
胡椒鹹雞......172

辣椒
星洲醃青辣椒......176
私房剁椒醬......180
私房剁椒醬蒸大魚頭......182

茄子
美極黑醋醃茄子......184

winter

陳皮
私房陳皮......188

泡菜
傳統廣東甜酸泡菜......190
梅醋醃蓮藕......192
醬油小黃瓜......194
酸辣千層大白菜......196

醬料
海南雞辣椒醬醃椰菜花......198
糟滷醉鮮鮑......200
沙爹XO醬......202
麻辣醃蘿蔔......204

味噌
味噌醃鹹蛋黃......206
味噌蛋黃軟殼蟹......208
味噌醃鱈魚......210

享受孤獨愛漫 • 遊......285
惜飲惜食......286

目 次
Contents

春 • spring

Kumquat • 金柑

214. Homemade salted kumquats
金柑の塩漬け

216. Steamed grey mullet with salted kumquats and fermented soybean paste
ボラの金柑塩漬と中国味噌蒸し

216. Duo kumquat marmalade
ダブル金柑ジャーム

217. Homemade kumquat wine
自家製金柑酒

Osmanthus kumquat wine
桂花金柑酒

Green plum • 梅子

218. Homemade green plum wine
自家漬け梅酒

219. Muscovado plum wine
濃醇黒糖梅酒

220. Green plum wine jelly
梅酒ゼリー

Green plum honey
蜂蜜梅干

221. Homemade green plum vinegar
自家製梅酢

222. Homemade muscovado green plum syrup
自家製黒糖梅シロップ

223. Homemade salted green plums
自家製梅干し

224. Steamed swimmer crabs with salted green plums
わたり蟹の梅蒸し

225. Braised chicken with young ginger and salted plums
梅、新しょうがと鶏の煮物

Strawberry • いちご

226. Strawberry jam
いちごジャム

228. Yoghurt with fruit preserve
ジャム入りヨーグルト

Candied strawberries in honey
いちごのはちみつシロップ

229. Zesty strawberry wine
情熱のいちごワイン

夏 • summer

Fruit wine • 果実酒

231. Homemade Chinese bayberry wine
自家製ヤマモモ酒

232. Guava wine
グアバ酒

Aged osmanthus Chinese bayberry wine
ヤマモモ桂花陳酒

233. Preserved fruit wine
ドライフルーツ酒

234. Japanese peach wine
日本のもも酒

Homemade raspberry wine
自家製ラズベリー酒

235. Black mulberry rum
黒桑の実のラム酒

236. Kyoho grape wine
日本巨峰酒

236. Japanese Ourin apple wine
日本の王林りんご酒

237. Homemade golden plum wine
自家製すもも酒

238. Cherry wine
サクランボ酒

Blackberry rum
ブラックベリーラム酒

239. Aged osmanthus lychee wine
ライチ桂花陳酒

Jam • ジャム

240. Passionfruit pineapple jam
パッションフルーツとパイナップルジャム

241. Rose peach jam
モモのバラ入りジャム

242. Honey kiwi jam
はちみつキウイジャム

243. Vanilla Mango jam
バニラマンゴージャム

244. Star-fruit muscovado jam
黒糖スターフルーツジャム

245. Aged osmanthus lychee wine jam
ライチの桂花陳酒入りジャム

Kyoho grape jam
プレミアム巨峰ジャム

246. Dried plum and pineapple jam
干し梅とパイナップルジャム

Ren Ren • 仁稔（ヤンニム）

247. The forgotten taste~
My Auntie's pickled Ren Ren
忘れられた味わい − おばの仁稔（ヤンニム）

248. Tofu and thousand-year egg cold appetizer dressed in Ren Ren juice
仁稔（ヤンニム）汁とピータン豆腐の和え物

Green Papaya • 青パパイヤ

249. Traditional sweet and sour green papaya pickles
伝統の青パパイヤ甘酢漬け

Passionfruit green papaya pickles
パッションフルーツとパパイヤの漬け物

Pickled vegetable • ピクルス

250. Acar (Nyonya assorted pickles)
アチャー（シンガポール　ピクルス）

Young ginger • 新しょうが

252. Pickled young ginger trio
新しょうが漬けトリオ

Winter melon • 冬瓜

254. Winter melon fermented soybean paste
冬瓜味噌

Steamed pomfret with winter melon fermented soybean paste
冬瓜味噌入りマナガツオ蒸し

255. Steamed pork patty with winter melon fermented soybean paste
冬瓜味噌入りひき肉蒸し

Green mango • 青マンゴー

256. Thai pickled green mango
タイ風の青マンゴー漬け

257. Stir-fried beef with Thai pickled green mangoes
タイ風の青マンゴー漬けと牛肉炒め

258. Green mango seafood salad in Thai style
青マンゴーのタイ風シーフードサラダ

White bitter melon • 白ゴーヤ

259. White bitter melon marinated in Okinawa Kurozatou
白ゴーヤの沖縄黒砂糖漬け

秋 • autumn

Apple • リンゴ

261. Homemade apple vinegar
自家製リンゴ酢

262. Pickled cherry tomatoes in apple vinegar
ミニトマトのリンゴ酢漬け

Cherry tomato vinegar jelly
ミニトマト酢のゼリー

263 Cherry tomato vinegar drink
ミニトマト酢のドリンク

Eggplant • ナス

Eggplant pickled with maggi's seasoning and balsamic vinegar
ナスのマギーと黒酢漬け

Tomato • トマト

264. Low-temperature air-dried tomatoes with truffle oil
ドライトマトのトリュフオイル入り

Cold tomato appetizer in plum dressing with mint leaves
冷やしミント、梅、トマト

Salted eggs • シエンタン

266. Homemade foolproof five-spice salted eggs
自家製香辛料と塩漬けの卵

267. Stir-fried bitter melon with five-spice salted eggs
香辛料塩漬け卵とゴーヤ炒め

Salmon • サーモン

268. Salmon Shoyuzuke with sake and dill
サーモンの清酒、ハーブ、醤油漬け

269. Pepper-scented salted salmon
サーモンのコショウ塩漬け

Peppercorn • コショウ

270. Pepper-scented salted chicken
鶏のコショウ入り塩漬け

Chilli • 唐辛子

271. Homemade chopped chilli sauce
自家製の唐辛子ソース

272. Steamed fish head with homemade chopped chilli sauce
自家製唐辛子ソース入りコクレンの頭蒸し

Singaporean pickled green chillies
シンガポール風の青唐辛子ピクルス

冬 • winter

Dried tangerine peel • 陳皮

273. Making your own dried tangerine peel
陳皮を自分で作る

Pickled vegetable • ピクルス

274. Traditional Cantonese pickles
伝統の広東ピクルス

275. Pickled lotus root in green plum vinegar
レンコンの梅酢漬け

Soy-pickled baby cucumbers
キュウリの醤油漬け

276. Sour and spicy Tianjin white cabbage pickles
酸辣(サンラー)の千層白菜

Sauce • 醤（ジャン）

277. Pickled cauliflower with chilli sauce for Hainan chicken
カリフラワーの海南鶏チリソース漬け

278. Wine-marinated abalones in distiller's grain sauce
糟滷（酒粕）酔っぱらいあわび

279. Satay XO sauce
サテーXO醤

280. Radish pickles with Sichuan peppercorns
大根の麻辣(まーらー)漬け

Miso • 味噌

281. Miso-salted egg yolk
アヒル卵黄の味噌漬け

282. Miso-salted egg yolk crusted soft-shell crabs
味噌卵黄のソフトシェルクラブ

283. Grilled cod in miso sauce
タラの味噌漬け

春 Spring

柑桔食療

柑桔可連皮食用，含有豐富的糖分及維生素C，根據中醫認為，柑桔性溫，味酸甘，具有理氣化痰、止咳、預防感冒及止渴解酒等食療效用。浸酒後飲用，不但氣味芬芳怡人，更有預防支氣管炎的作用，實在是價廉物美的養生健康佳品。

醃製了十年的鹹柑桔

過年後
齊齊醃
鹹柑桔

其實，一些經常在你日常生活中會見到的食材，往往都很容易被人忽略了它的食用價值及美味，柑桔便是一個好例子。過完年，很多人家裏都可能有一兩盆柑桔，擺設完了，就被當垃圾般被棄置，實在很可惜。我們可以把它摘下來，或向隣居朋友收集一些回來，洗乾淨，醃製成鹹柑桔後，它馬上便會變成你家中的「寶貝」！

醃製一年後的鹹柑桔基本上已經可以食用了，有清熱、下火、利咽、健胃及幫助消化的食療功效。在喉嚨痛、聲音沙啞、用聲過度的時候，以兩粒鹹柑桔壓爛，加入上等蜂蜜，以溫水沖飲。這是一道家傳戶曉的民間靈方，如是我自己飲的話，還會加半粒新鮮青檸汁，味道更加清香味美！除此之外，你還可以將鹹柑桔拌以七喜、薑啤、梳打水沖飲，解渴怡神，各俱風味！(建議每次兩粒，壓爛)。

醃鹹柑桔

金柑の塩漬け...214
Homemade Salted Kumquats...214

以同樣方法，一樣可以醃製出鹹檸檬。

❀材料
新鮮柑桔 約900克
粗鹽 約900克
蒸餾水或白醋 1/4 杯

❀做法

1. 新鮮柑桔沖洗乾淨，瀝乾水分。在太陽下曝曬兩至三天，至柑桔表皮開始收水，出現微微皺紋，去蒂及用布將柑桔抹乾淨。

2. 取一可密封、寬口及透明的玻璃容器，重複交替地先放一層粗鹽，再放入一層柑桔，最後加入約1/4杯蒸餾水或白醋。

3. 待容器裝滿，然後封密，寫上醃製日期，放置在陰涼、乾燥及陽光照射不到的地方(建議最好放在廚櫃內)。約一星期後，柑桔便會慢慢滲出水分，粗鹽亦隨着慢慢溶解，而柑桔亦會由最初鮮艷的橙黃色，慢慢變成淺啡至深啡色。

釀製半年後的鹹柑桔。

醃製約一年後的鹹柑桔。這個時候鹹柑桔就可以用啦!

貼士 ✿

1. 醃製柑桔之前,將柑桔放在陽光下曝曬幾天的原因,是一來可利用太陽的紫外線殺菌,二來可將柑桔表皮的水分收乾,這樣醃製出來的鹹柑桔便不易變壞,更可長時間保存。三來柑桔皮收乾水分後,柑桔果的味道會濃些,醃製後所散發出來的柑桔香氣亦會強些!而加入少許蒸餾水或白醋是可以加速誘導鹽分溶解及有足夠水分「養」着鹹柑桔。切勿加入生水,因為未經煮沸的生水含有微菌,容易令食物發霉。

2. 凡是醃製食物,宜選用寬口及透明的玻璃容器。因為醃製食物必須要易於取放,隨時看到瓶內食材的變化及可長時間存放。高濃度鹽分及酸性的醃製食物都不適宜使用塑膠容器。容器泡醃食物之前,一定要用滾水燙過,吹乾後才使用,以達消毒作用。

3. 醃製柑桔前,一定要確保柑桔完全乾身,不可沾有生水,否則很易發霉。

4. 醃製好的鹹柑桔,因含有高濃度的鹽分,可存放很長時間,就像陳年舊酒般,存放時間越長,醃製效果及味道就越醇厚,有人甚至將它存放幾十年之久,像名酒般看待!

5. 很多用作過年擺設的柑桔,花農都為了使它看起來更亮麗及可更長時間擺放。都在柑桔上噴灑一些化學物品以達以上效果,所以大家一定要將柑桔徹底清洗乾淨後才醃製。或購買些在街市生果檔一籮蘿大量售賣,不是用作觀賞的柑桔較為安全。

鹹柑桔麵豉醬蒸烏頭

ボラの金柑塩漬と中国味噌蒸し 216

Steamed grey mullet with salted kumquats and fermented soybean paste 216

🍀 材料

烏頭 一條(約900克)
鹹柑桔 六至八粒
麵豉醬 一湯匙
薑片 三至四片
油 兩湯匙

🍀 做法

1. 烏頭劏好洗淨，薑片放底，放上烏頭；鹹柑桔切片鋪在烏頭上，並均勻放上麵豉醬。
2. 水滾後大火蒸約八分鐘，取出，將油煮滾淋上魚身即成。

貼士 🍀

• 蒸魚時間視乎魚之大小而適當加減，亦可用同樣方法蒸其他魚類。我也試過蒸排骨、肉眼筋，加些河粉或陳村粉放底……美味極了！

春 _ Spring

金銀柑桔果醬

ダブル金柑ジャーム...216
Duo kumquat marmalade...216

醃製好的鹹柑桔，除了沖飲之外，我還將它混合新鮮的柑桔做成果醬，酸甜之間，夾雜着鹹柑桔的獨特鹹香味道，互相調和，份外怡神。

🍀材料

鹹柑桔　150克(約16粒)
新鮮柑桔　1公斤
砂糖　200克
青檸檬汁　2湯匙

蜂蜜　1/2杯
水　800毫升

🍀做法

1. 鹹柑桔切碎去籽，新鮮柑桔洗淨瀝乾水分，切碎去籽，與砂糖、青檸檬汁撈勻，放置一晚讓它出水。

2. 第二天加入水，大火煮滾後改小火，邊煮邊不時攪動材料，煮約20分鐘至汁液開始起泡，變得亮澤及已變成果醬的濃稠狀態便可關火，將果醬盛起，涼卻後便可入已消毒處理的容器封好保存，慢慢享用。

貼士 🍀

● 一定要將柑桔的籽去掉，否則果醬味道會帶有苦味及吃果醬時會咬到種子破壞口感。

私房柑桔酒

自家製金柑酒...217
Homemade kumquat wine...217

材料

新鮮柑桔 3.375公斤
廣東米酒 2.7公斤
冰糖 1.8公斤

做法

柑桔洗淨，去蒂，抹乾水分，與米酒、冰糖放入一可密封的玻璃瓶內，封密，存放在陰涼、乾燥及陽光照射不到的地方，約一年後便可飲用。

桂花柑桔酒

桂花金柑酒...217
Osmanthus kumquat wine...217

🍀 **材料**
新鮮柑桔　1.575公斤
桂花陳酒　3樽（每枝750毫升）

🍀 **做法**
　柑桔洗淨，去蒂，抹乾水分
　與桂花陳酒放入可密封的玻
　璃瓶內，封密，存放在陰涼
　乾燥及陽光照射不到的地
　方，約半年後便可飲用。

又見梅子成熟時！

嗯⋯好誘人

每年四至五月，街市的菜檔都多了一籮籮青綠色的梅子，我知道，又是泡製梅酒的時候了。通常當我在街市大量入貨的時候，旁邊便開始有人七嘴八舌，不停地問：「梅子用嚟點食㗎？浸酒㗎？點浸㗎？」在這本書，我一於教教大家怎樣去浸泡自家梅子酒啦！

其實，看見這些圓潤可愛的梅子，很多人都很感興趣，但又不知如何處理。其實，這些每磅才十多元，咬下去又酸又澀的青梅，是可以做出很多美味的食物，最普遍如泡浸梅子酒、醃製成各種不同味道的脆梅、酸梅、鹹梅、泡醃成梅子糖漿、梅子醋等等，然後再用這些製成品，變化出各種不同的食品或飲料。

這些「手作仔」，在台灣及日本，是很多家庭主婦駕輕就熟的手藝，這些醃製好的成品，還可以長時間保存而且愈舊愈醇，愈陳愈香。一瓶親手泡製的梅酒，就如多年的老朋友，相處時間愈長，感情就愈深厚。當你打開一瓶陳年梅酒招待至愛親朋時，誘人心扉的香氣瀰漫滿室。

此時，你會發現，泡釀梅酒時的漫長等待，是多麼的值得啊！

自家漬け梅酒 ..218
Homemade green plum wine ..218

浸泡私房梅子酒

全部材料一起泡浸，梅子變皺縮小。

先放梅子、酒，後放冰糖，梅子圓潤可愛。

材料

青梅子　6磅
冰糖　5磅
米酒　6磅

做法

1. 梅子沖洗乾淨，瀝乾水分，讓它自然吹至完全乾身，然後用竹籤挑去梅子的蒂，再在梅子卜刺一至兩個洞，使梅液易於滲出。

2. 取一可密封、寬口及透明的玻璃容器，分別放入梅子、冰糖及米酒。

3. 然後封密，並寫上泡浸日期，放置在陰涼、乾燥及陽光照射不到的地方，偶爾將瓶子搖晃一下，待冰糖完全溶化之後，便可讓它靜止存放，約三個月後，就會變成一瓶金黃通透、清香迷人、酸甜可口的自家泡製梅子酒了。

泡釀梅酒貼士

1. 雖然梅子酒泡浸約3個月便可以享用，但是梅酒存放愈久，發酵成熟的風味愈發誘人，味道愈見馥郁醇厚，讓人急不及待地想去品嘗。

2. 一般泡浸果實酒，宜選用30度以上的米酒，因為高濃度酒精更能誘發果實的芳香味道。玉冰燒、廣西米酒、雙蒸、三蒸、九江米酒甚至日本清酒均可。

3. 泡浸梅酒時（尤其是大量製作），為了方便，你可以一次過將梅子、冰糖及酒一齊放進瓶子泡浸，但當冰糖完全溶化後，梅子裏的水分會因為酒中含高濃度的糖分而被慢慢抽出，令梅子變皺縮小不太好看。另一做法是先放入梅子及酒，浸泡兩至三個月後，讓梅子充分吸收酒精後，才放入冰糖。此時梅子內水分的濃度已與冰糖溶化後的酒達至平衡，泡完的梅子仍保持圓渾飽滿，非常可愛。無論哪一種做法，完成的梅酒味道都沒分別。

4. 選用較黃及熟的梅子，泡出來的梅酒顏色會較深，香味會較濃；若選用較青及硬身的梅，泡出來的梅酒顏色會較清，但梅子的酸味會較重。

5. 梅子酒不但好飲，還可調校各種特色飲品及製造各種甜品。適量飲用，據說還有祛風濕、健脾開胃、消除疲勞等食療功效。

6. 容器適宜用寬口及透明的玻璃容器，因為醃製食物必須要易於取放，隨時看到瓶內食材的變化及可長時間存放，高濃度鹽分及酸性的醃製食物都不適宜用塑膠容器。使用容器泡醃食物前，一定要用滾水燙過、吹乾，以達消毒作用。

香濃黑糖梅酒

泡浸梅酒除了使用冰糖外，還可以選用黑糖，黑糖那股獨有的焦糖香氣濃厚馥郁。泡浸好的黑糖梅酒除了飲用之外，更是用來炮製甜品的上好材料。

濃醇黑糖梅酒...219
Muscovado plum wine...219

❦材料

半熟或熟梅子　6磅
米酒　6磅
黑砂糖　5磅

❦做法

1. 梅子沖洗乾淨，瀝乾水分，讓它自然吹至完全乾身，然後用竹籤挑去梅子的蒂，再在梅子上刺1-2個小洞，使梅液易於滲出。

2. 取一可密封、寬口及透明的玻璃容器，放入梅子、米酒及黑砂糖，然後封密，並寫上泡浸日期，放置在陰涼、乾燥及陽光照射不到的地方，偶爾將容器搖晃一下，待黑砂糖完全溶解之後，便讓它靜止擺放約3個月後便可以飲用了。

貼士❦

1. 黑糖梅酒取其香甜口味，而且無論黑糖梅酒本身或泡浸完梅酒的那顆充滿黑糖風味的梅子，都很適合用來配襯各類甜品。宜選用半熟或已變黃的梅子，熟梅與黑糖結合之後，會令完成後的梅酒無論在味道及香氣上都更加濃郁豐厚。

2. 你亦可以選用傳統的黑糖泡浸，如著名的日本沖繩黑糖，味道還會更香、更濃，但完成品會較渾濁；如用黑砂糖的話，泡浸出來的梅酒會清澈一些，但味道就不及傳統黑糖般香濃了。

梅酒凍

梅酒ゼリー...220
Green plum wine jelly...220

泡浸好的梅酒除了直接飲用外，亦可做成各種精美甜品，梅酒凍是其中一款既簡易又美味的選擇。清香，酸甜，帶點懶洋洋的醉意，是夏日或晚餐後的寫意甜品。

❧材料

自製梅酒　1公升
魚膠片　50克(約10片)

❧做法

1. 魚膠片先用冰水浸約10分鐘使其軟身，然後揸乾水分備用。
2. 梅酒加熱至60-70℃左右，放入魚膠片後攪拌至完全溶解，再用密篩過濾以去除雜質及未溶解的魚膠片殘渣。
3. 將梅酒倒入預備好的甜品容器或盤中，涼卻後放入雪櫃冷凍凝固成啫喱狀即成。

貼士❧

1. 梅酒凍可直接食用，亦可切成自己喜愛的形狀配以各類甜品或飲品，倍感涼快。
2. 可隨自己喜歡梅酒凍的軟硬度而增加或減少魚膠片的份量。

在春天，市場上處處有梅子的踪影！

呷醋有益

近年坊間很流行飲用水果醋，所謂水果醋，就是以水果用醋泡浸，經長時間醞釀而成。現在市面上均有各式各樣的果醋產品，而根據不同水果本身的食療作用混合了醋之後，據講都有不同的食療作用，例如最流行的蘋果醋，據說便有防止脂肪積聚及幫助消除鼻敏感的作用，亦由於可口，深受一眾追求健康飲食人士的歡迎！

而醋本身的確很有食療功用，例如：

*促進口水、胃液的分泌，調整腸胃機能，幫助消化

*使血液變得潔淨，促進脂肪、糖分分解

*幫助排出體內多餘的膽固醇

*增加皮膚、頭髮的亮澤，從血達到美容作用

*用醋來烹調菜式，更加是美味開胃

而用梅子泡浸出來的梅子醋，更兼備了梅子及醋兩方面的食療效用，可謂一舉兩得。梅醋不單止好飲，還可以止渴、消除疲勞、解熱、止咳、健胃，更可以有效中和酸性及鹼性的食物。當你食慾不振的時候，飲一小杯梅醋，包你胃口大開。一般來說，梅醋可以直接飲用，每次約一小杯，若嫌太酸的話，可調以適量開水或蜂蜜飲用，你更可以運用你烹調上的創作力，用梅醋代替其他醋來泡醃各類蔬果泡菜，如梅醋苦瓜、梅醋石榴、梅醋蘋果等……加些橄欖油拌沙律，清香味美，甚至燜煮及烹調各類菜式，都非常之健康有益。

最重要的是，要成功製作梅醋，其實真的易如反掌，只是需要一點時間及耐性去等待而已。

剛泡浸的梅醋

泡浸一年後的梅醋

自家製梅酢 ...221
Homemade green plum vinegar ...221

❧材料

半熟或熟梅子　1.5公斤
白醋　1.5公升
冰糖　800克

❧做法

1. 將梅子洗淨，徹底吹乾水分，用牙籤挑去梅子的蒂，再在梅子上刺入一至兩個小孔，使梅液更易滲出。

2. 容器放入梅子，倒入白醋及冰糖，將容器封密，放在陰涼及陽光照射不到的地方，偶爾將容器搖晃一下，待冰糖完全溶解後，讓它靜止擺放，約六個月後，梅醋便可以飲用了。

貼士 ❧

1. 因為梅子與醋都是酸的，因此要選用較熟變黃的梅子 (若買回來的梅子太生，可以不要清洗，將梅子放置三、四天後，梅子自然便會熟成變黃)，選用較生及青色的梅了泡出來的梅醋味道會酸些，選用熟梅泡出來的梅醋不會那麼酸，而且梅子的味道及香氣更會甘香濃郁些。

2. 泡浸梅醋一定要加上糖分，以中和它的酸性，除了冰糖之外，你還可以用麥芽糖甚至黑糖代替，各具風味，而份量可以因應個人的口味而加多或減少，甚至泡浸完成後你還是覺得太酸的話，你還可以在調配飲用時適量添加自己喜歡的糖類。

3. 由於梅醋是可以直接飲用，所以泡浸梅醋時宜選用較為優質的進口白醋，這些白醋醇度較高，沒有醋精，泡浸完成後直接飲用也不會覺得嗆喉，而一般國產廚用白醋，只適宜調味、烹煮食物及醃製泡菜。

梅子蜂蜜

蜂蜜梅干...220
Green plum honey...220

甜美的蜂蜜吸收了梅子的酸香風味，確實是夏天一道解渴生津的養生妙品！梅子在四、五月當造的時間，正是醞釀梅子蜂蜜的最佳時候，當你看見荔枝登場，見到人們向捲着熱浪的海灘進發的時候，亦代表梅子蜂蜜已經醞釀成熟，這便是享受成果的時候了！

🍀 材料
半熟或青梅　1公斤
蜂蜜　1.5公斤

🍀 做法
1. 梅子沖洗乾淨，瀝乾水分，然後用竹籤挑去梅子的蒂，再在梅子上刺一至兩個小洞，使梅液易於滲出。
2. 將梅子放入容器，倒入蜂蜜，蓋好，放在陰涼、乾燥及陽光照射不到的地方，約三個月至梅子汁液排出，與蜂蜜融合，同時梅子亦會縮小變皺，這時便可以飲用了。

貼士 🍀

1. 做梅子蜂蜜不宜選用太熟的梅子，因為熟梅的氣味較強，而且水分較多，梅汁與蜂蜜結合之後會搶去蜂蜜的香氣。
2. 蜂蜜因為吸收了梅子排滲出來的梅液而稀釋了，完成後的梅子蜂蜜會較原來的狀態稀，這是正常現象。
3. 梅子蜂蜜可以直接加水稀釋飲用，冰凍後更添風味，但不宜熱飲，因為高熱會破壞蜂蜜原有的天然營養價值；它可以配不同味道的雪糕、甜品或糕餅，美味怡神。

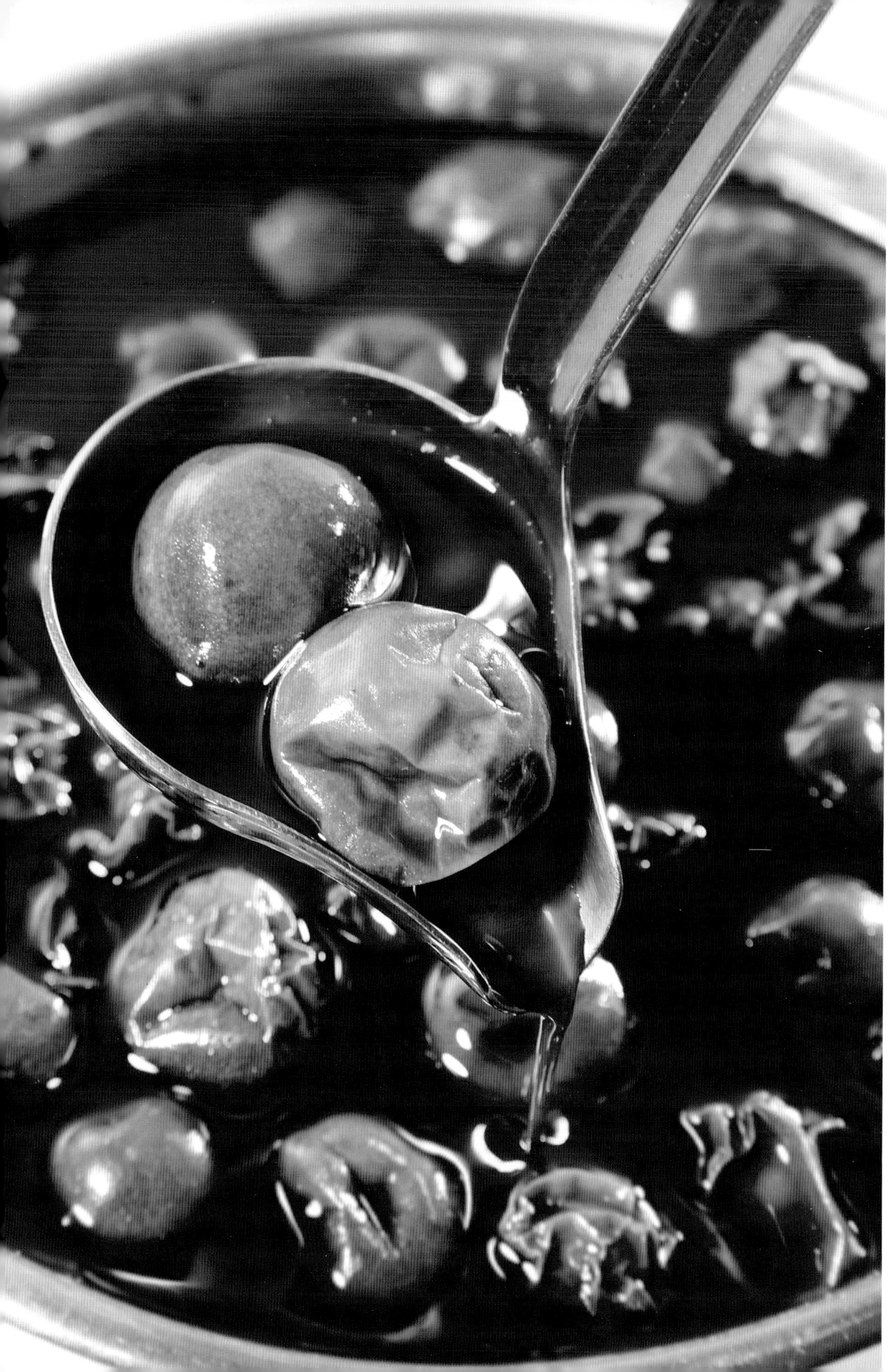

自家製黑糖梅シロップ…222
Homemade muscovado green plum syrup…222

梅子經黑糖醃製後，黑糖吸收了梅子的精華變成香氣馥郁的濃縮液，加入果汁或梳打水，便是一杯解渴怡神的夏日飲品。梅子糖漿亦可淋在各式的雪糕或甜品上，亦可用它做成各種甜品、果凍或蛋糕。而那顆金黃色的梅子，經雪凍後，就已是妙不可言的甜品！

材料

半熟梅子　2公斤
黑砂糖　2公斤
米醋　半杯
粗鹽　3茶匙

做法

1. 梅子沖洗乾淨，瀝乾水分，然後用竹籤挑去梅子的蒂，再在梅子上刺一至兩個小洞，使梅液易於滲出。
2. 取一大盆，重複交替地先放一層黑砂糖，再放一層梅子。
3. 直至裝滿大盆，但最後一層必須是黑砂糖。
4. 加入粗鹽。
5. 用手輕輕將梅子及黑砂糖壓實，使梅液滲出時更易接觸到砂糖，加速溶化。
6. 淋上米醋。
7. 用保鮮紙將大盆封密，並寫上醃製日期。
8. 約一星期後，梅子慢慢滲出汁液，變成淡黃色，黑砂糖亦隨着慢慢溶解，此時，要將梅子與黑砂糖撈勻。
9. 約兩星期後，黑砂糖因混合梅子汁液而繼續溶化，成為黑糖漿泡浸着梅子，梅子亦開始變軟及成金黃色。
10. 此時可將梅子及黑糖漿舀入一可密封、寬口及透明的玻璃容器內，然後封密，寫上日期，放入雪櫃，讓梅子再繼續熟成。
11. 約兩個月後，梅液已將黑砂糖完全溶化，成為黑糖梅子漿。如梅酒一樣，它散發出誘人的梅子香氣，但製作過程就不需要像梅酒那樣，經過漫長的等待時間。

用白砂糖做的私房梅子糖漿

貼士 🍀

1. 加入米醋的目的是引導糖分加速溶化及增強保存性。

2. 除了用黑砂糖之外，亦可以使用普通的白砂糖。黑砂糖以日本沖繩或台灣出產的為佳，味道較濃厚香甜，而白砂糖的醃製效果則較清香。

私房鹹水梅

自家製梅干し...223
Homemade salted green plums...223

鹹水梅是一樣很普及的廣東家庭菜原材料，一般用於燜、蒸及煮的菜式，梅子燜鴨、梅子蒸排骨你一定吃過吧，但若自己親手造一瓶鹹水梅做菜，那便一定更加好味了。

🦋材料

熟梅子　1.2公斤
粗鹽　　1.2公斤
白醋　　半杯

🦋做法

1. 梅子沖洗乾淨，瀝乾水分，然後用竹籤挑去梅子的蒂。
2. 取一已消毒的容器，重複交替地放入一層粗鹽，再放一層梅子，但最後一層必須要是粗鹽，倒入白醋，封密，貼上醃製日期，放在陰涼、乾燥及陽光照射不到的地方，大約6個月待粗鹽完全溶解、梅子全部變成啡色及梅子全部變皺便可以使用了。

貼士 🦋

1. 醃製時加入白醋，目的是引導鹽分加速溶化，不會影響味道。
2. 醃製完成的鹹水梅可以用來做成各款美味菜式的調味料，或用以調配各類飲品，如梅子蜜、梅子七喜、梅子梳打等等。

梅子蒸花蟹

わたり蟹の梅蒸し...224
Steamed swimmer crabs with salted green plums...224

♣ 材料
花蟹　2隻(每隻約600克)
自製鹹水梅　約10粒
橄欖油　3-4湯匙
葱花　適量

♣ 調味料
水　2湯匙
米酒　2湯匙
番茄醬　3湯匙
梅子醬　1湯匙
砂糖　1 1/2湯匙
生抽　1/2茶匙

♣ 做法
1. 蟹劏好洗淨，斬件，放於蒸碟上。
2. 鹹水梅去核後搗爛，與調味料攪勻。燒熱橄欖油，下調味料炒成蒸蟹梅子醬，然後均勻放於蟹上，水滾後大火蒸約8-10分鐘，取出灑上適量葱花即成。

貼士 ♣
1. 除了蒸花蟹，梅子醬用來蒸蝦或其他貝殼類海產也非常合適。
2. 鹹水梅的核其實也很有味道，若不介意賣相有影響的話，蒸蟹時連同鹹水梅的核同蒸會更加出味。

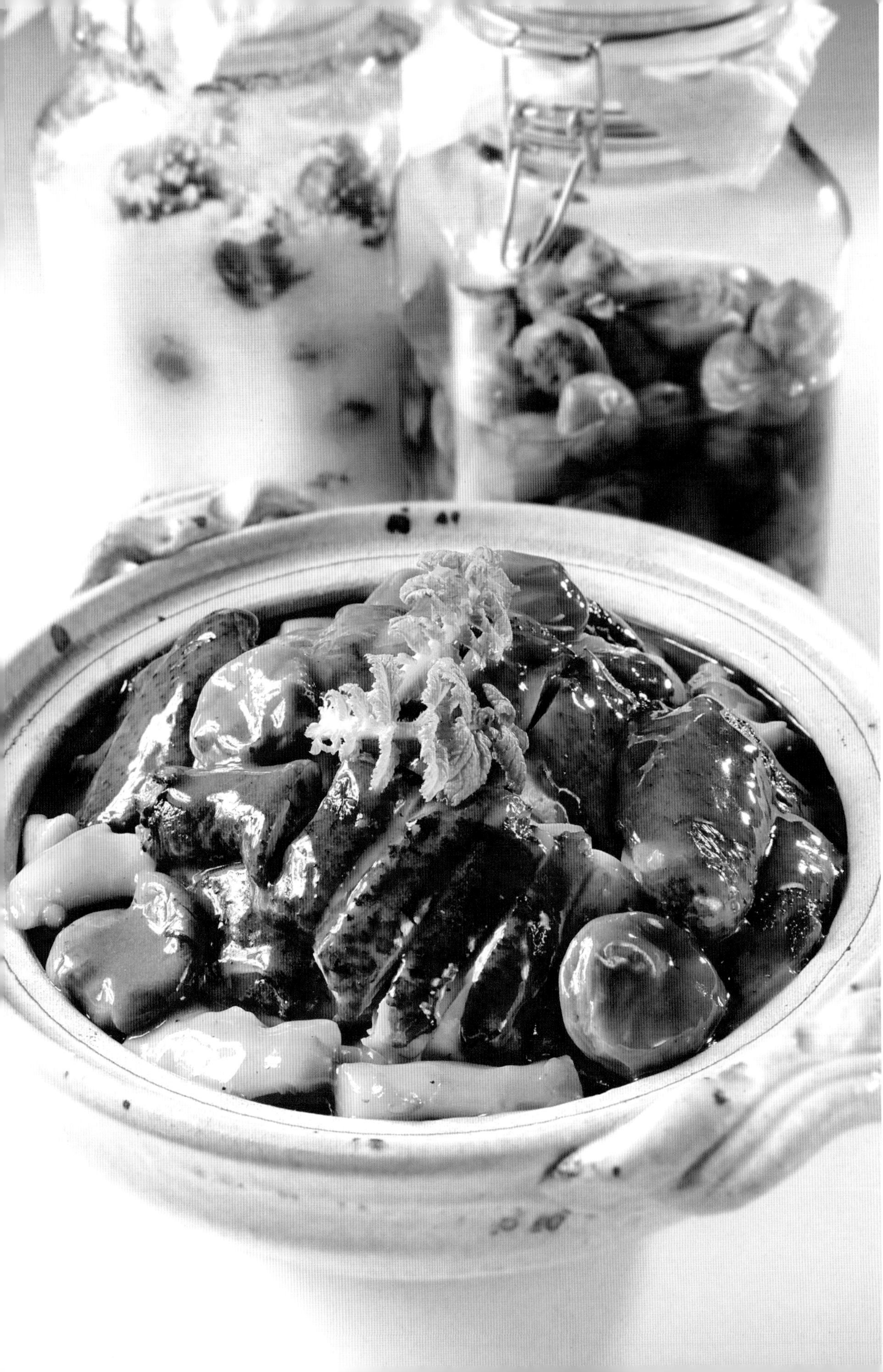

梅、新しょうがと鶏の煮物...225

Braised chicken with young ginger
and salted plums...225

❦ 材料

鮮雞　1隻(約1.5公斤)
新鮮子薑　50克
糖醋子薑　100克
米酒　少許
老抽　約1湯匙

❦ 釀料

鹹水梅　80克(約10粒)
蒜頭　3粒
碎冰糖　40克
磨豉醬　1湯匙

❦ 燜料

生抽、老抽　各1湯匙
八角　1粒
雞湯　約1公升

❦ 做法

1. 先將新鮮子薑洗淨，拍鬆，用粗鹽約1/2茶匙撈勻醃約半小時；鹹水梅去核(核留用)，梅肉與蒜頭、冰糖一起舂碎，然後與磨豉醬拌勻備用。

2. 雞洗淨，吸乾水分，用醃料塗勻全身，醃約30分鐘，將釀料及留起的梅子核用少許油炒勻至冰糖完全溶解，然後放入雞肚內，用廚用粗針縫密。

3. 鑊燒熱油約1湯匙，爆香新鮮子薑，潷入少許米酒，加入燜料，將雞放入，加蓋，大火煮滾後改慢火燜約15分鐘，將雞反轉再燜約15分鐘至雞熟透。

4. 撈起雞及子薑，隔去雞汁內的油分，將雞肚內的釀料及糖醋子薑加入雞汁內，大火收乾至約1量杯的分量，用隔篩隔去汁內的雜質，再用生粉水埋一個薄獻。

5. 將雞斬件，將雞汁淋在雞上，再將剛才撈起的子薑放在雞旁伴吃，綴上適當裝飾即可亨用。

貼士 ❦ ▬▬▬▬▬▬▬▬▬▬▬▬▬▬▬▬▬▬▬▬▬▬

1. 燜雞時一定要用慢火，這樣雞肉才會嫩滑。

2. 此道菜式可以變化成手撕雞，然後將撕出來的雞肉放在蒸熱，混有臘味、菜脯、炒花生、芫茜及葱花的糯米飯上，再淋些梅子雞汁伴吃，是一道非常滋味暖胃、適合在冬天享用的菜式。

3. 糖醋子薑的做法，請參閱第125頁。

草莓果醬

いちごジャム ...227
Strawberry jam ...226

草莓，
就像愛情…
它擁有一種甜蜜
而迷人的香氣，
讓人歡喜若狂的
動人氣息。

♣ 材料

草莓　1千克
砂糖　200克
鹽　1/2茶匙
青檸汁　1個份量

♣ 做法

1. 草莓洗淨，抹乾水分，去葉及蒂，切細粒，與砂糖、鹽、青檸汁撈勻，用保鮮紙封好，放入雪櫃，置放一天，讓它出水。

2. 第二天將草莓連同排出的水分倒入鍋中，以小火煮滾，撈出多餘的泡沫，盛起待涼，再放入雪櫃存放一天。

3. 第三天將草莓用篩隔起，將排出的汁液倒入鍋中，以大火煮滾後改小火煮至汁液開始濃稠(約十分鐘)，然後將先前隔起的草莓回鍋拌勻，以小火繼續煮約五分鐘，此時汁液開始起泡，草莓亦開始變得光亮及已變成果醬的濃稠狀態，這時果醬基本上已經完成，你便可因應個人喜歡果醬的濃稠程度而選擇關火，這樣果醬會汁液多些，或再煮長一點時間而濃稠一些。

4. 將果醬盛起，趁熱裝入已經消毒處理的容器內，封好保存，可慢慢享用了。

＊ 跟隨以上的方法，可製成果肉和汁液比較分明的草莓果醬。

貼士 ❦

1. 煮滾汁液的期間會有一些泡沫浮渣，要將它撈起，這樣做出來的果醬便會晶瑩亮麗，美觀吸引。
2. 建議選用美國出產的長椗草莓，因為其品種無論香氣及味道都非常濃郁，做出來的果醬更香氣四溢，充滿誘惑！
3. 完成後的草莓果醬可以配麵包、各類糕餅、乳酪、雪糕等等⋯⋯非常美味。

另類口味

傳統草莓果醬

用以上方法做出來的果醬效果是汁液與果肉比較分明，汁液色澤鮮色亮麗，果肉亦比較有口感，若想果肉與汁液混合得濃稠一些，像傳統果醬一般果肉質地較「醬」的話，可以將做法步驟(2)的草莓及排出的汁液直接倒入鍋中，以大火煮滾後改小火，煮至果肉與汁液起泡及變成濃稠狀，便成傳統果醬般的狀態了。

黑糖草莓果醬

可將白砂糖改用黑糖，這樣做出來的草莓果醬又別有一番風味。

原粒草莓果醬

你更可選用日本或韓國體積較小品種的草莓來以做法1、2、3的方法來做果醬，完成後原粒入樽，賣相非常可愛，用來做蛋糕或直接用來做其他甜品的配料，簡直是無懈可擊。

傳統草莓果醬

原粒草莓果醬

ジャム入りヨーグルト...228
Yoghurt with fruit preserve...228

將不同風味的果醬配以新鮮乳酪，是一道感覺清新而又味道天然的健康早餐，如果每天可以享用自己親自烹調出來的果醬，原來，幸福是可以自己製造的！

❀材料

新鮮原味乳酪　約半杯
果醬　滴量

❀做法

將適量果醬舀入乳酪內，
吃時將兩者拌勻便可。

貼士❀

開啟後的果醬必須放入雪櫃貯存。

いちごのはちみつシロップ ..228
Candied strawberries in honey ..228

材料

日本或韓國小粒品種草莓　1.5公斤
蜂蜜　1公斤

做法

將草莓洗淨，去蒂，用廚紙徹底吸乾水分，放入容器內，倒入蜂蜜，蓋好，放入雪櫃約一個月，至草莓水分排出及體積縮小變皺即成。

貼士

1. 置放期間草莓會排出汁液，但因為濃度較稀會浮在上層，所以需要大概一星期攪拌一次，讓草莓汁液與蜂蜜能均勻溶合，這樣做出來的蜂蜜會更充滿草莓的香氣及風味。

2. 完成後的草莓蜂蜜可以像平時飲用蜂蜜般直接用水稀釋後飲用，冷飲熱飲均各俱風味，或可淋於各類雪糕甜品糕餅，美味滋潤。

3. 選用體積較小的日本或韓國品種草莓，是因為水分更易於排出，其香氣較濃，色澤亦較鮮紅亮麗，成品的賣相會更吸引。

4. 不要將草莓切開來泡浸，因為這樣泡浸出來的蜂蜜會很渾濁，雜質較多，不美觀。

情熱のいちごワイン...229
Zesty strawberry wine...229

說它催情，是因為它非常性感艷麗的玫瑰紅色，加上它那股充滿誘惑的迷人香氣，怎不讓人神魂顛倒、胡思亂想！

🌸 材料

美國長桎草莓　2.7公斤
泰國青檸檬　2個
米酒　3公升
冰糖　1.35公斤

🌸 做法

1. 草莓洗淨，去桎，瀝乾水分後用廚紙徹底吸乾水分，再放在筲箕內放置約1-2小時，以確保水分完全吹乾；青檸檬洗淨，抹乾水分，一開為二切開，挑去種子。

2. 取一可密封、寬口及透明的玻璃容器，放入草莓、青檸、米酒及冰糖，然後封密，並寫上泡浸日期，放置在陰涼、乾燥及陽光照射不到的地方，偶爾將容器搖晃一下，待冰糖完全溶解之後，便可以讓它靜止擺放約6個月後便可以飲用了。

貼士 🌸

1. 加入青檸檬的作用是中和草莓的甜味，其次亦可將草莓的甜香味道更加得以提升。

2. 冰糖的份量可因應每人喜歡的甜度而作調整。

3. 建議選用美國出產的長桎草莓品種，因為其香氣及味道都較其他品種更為濃郁，泡浸出來的果酒效果便更加理想。

Summer

夏人

偷得浮生半日閒，微醺，最悠然。

果酒是水果與酒經過長時間浸泡，

酒精將水果的味道吸取出來，

同時讓酒吸收果實的香氣及味道，

成為屬於該有果實味道的果酒。

不同的果實配以不同的酒泡浸出來的果酒，

各有不同的味道及風味，

這亦是泡浸果酒吸引人之處。

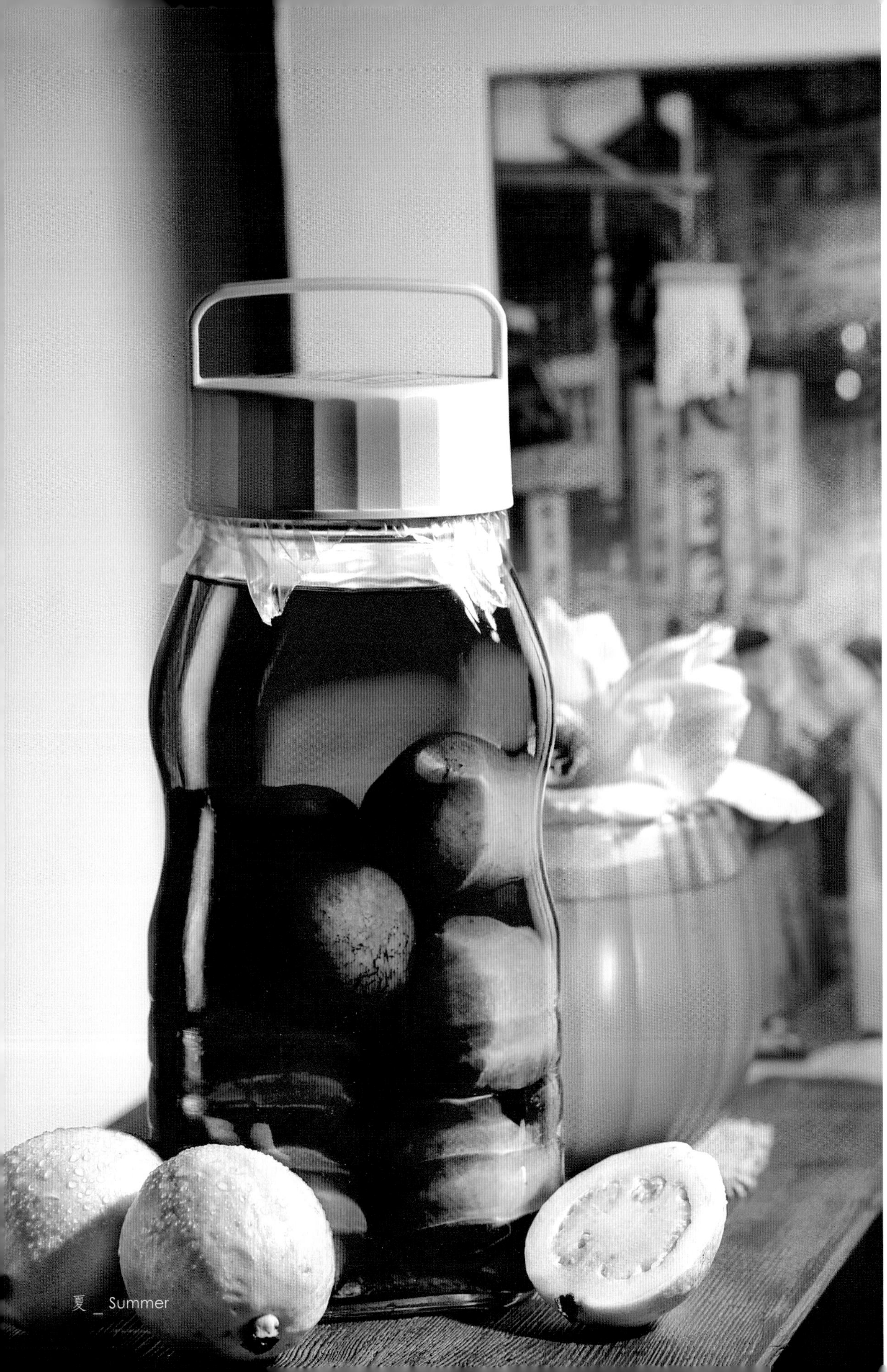

グアバ酒 232
Guava wine 232

番石榴，我們佛山人稱之為「雞屎果」。我們小時候住的地方在後園便有一棵非常高大的「雞屎果」樹，表弟還經常因為偷吃過度而常常便秘！它的香味非常奇特，與熱情果的香味非常相似，而且很濃烈，只要放一盆在飯廳，便可整間屋都可聞得到。我很喜歡這種味道，用它泡浸出來的果酒更是芳香撩人，滿室生香。

❀ 材料

本地番石榴　800克
三蒸米酒　1公升
冰糖　250克

❀ 做法

1. 番石榴洗淨，抹乾水分，一開為二切開。
2. 取一可密封、寬口及透明的玻璃容器，放入番石榴、三蒸米酒及冰糖，然後封密，並寫上泡浸日期，放置在陰涼、乾燥及陽光照射不到的地方，偶爾將容器搖晃一下，待冰糖完全溶解之後，便可以讓它靜止擺放約三個月後便可以飲用了。

貼士 ❀

1. 宜選用切開後果肉是粉紅色，俗稱胭脂紅的本地番石榴。因為此品種番石榴的香味特別濃烈誘人，用它來泡浸番石榴酒便史加香氣逼人了。
2. 要挑揀生熟適中的番石榴才可泡浸出美味怡人的番石榴酒，因為太生的不夠香，太熟的果肉太腍，經長時間泡浸後容易分解，令果酒渾濁。

自家製ヤマモモ酒...231
Homemade Chinese bayberry wine...231

高濃度的酒精更
能誘發果實的芳
香味道，這瓶用
高粱泡浸的楊梅
酒，無論香氣、
味道都非常芬芳
醇厚。

❧ 材料

新鮮楊梅　1.35公斤
高粱酒　1.35公斤(55度)
冰糖　675克
米酒　適量（消毒楊梅用，
要蓋過楊梅）

❧ 做法

1. 楊梅洗淨，完全瀝乾水分後，先用米酒泡醃6-8小時，再用篩隔起楊梅，瀝乾米酒備用。

2. 取一可密封、寬口及透明的玻璃容器，分別放入楊梅、冰糖及高粱酒，然後封密，並寫上泡浸日期，放置在陰涼、乾燥及陽光照射不到的地方，偶爾將容器搖晃一下，待冰糖完全溶化之後，便可以讓它靜止擺放約6個月後便可以飲用了。

貼士 ❧

1. 由於楊梅沒有果皮包裹着，而且楊梅果肉的質感是一梳梳充滿水分，在整個生長過程中都完全暴露於生長環境空間，容易在果肉內滋生昆蟲及微菌；而泡浸果酒是須要長時間存放，因此，在泡浸楊梅酒之前，一定費用30度以上的米酒浸泡半天，以達消毒作用，這樣泡浸出來的楊梅酒才能保存更長時間及衛生安全。

2. 平時亦應先用鹽水泡浸楊梅約30分鐘才進食，一來以達消毒作用，二來像吃西瓜的原理一樣，會令楊梅的味道更加甜美。

楊梅桂花陳酒

ヤマモモ桂花陳酒 232

Aged osmanthus Chinese bayberry wine 232

桂花陳酒有一股難以言喻的撲鼻幽香,我覺得它可配襯各類食材,尤其是水果,都能將它的獨特氣味表露無遺。

❀ 材料

新鮮楊梅 1.35公斤
桂花陳酒 3樽 (每樽750毫升)
米酒 適量(消毒楊梅用,要可蓋過楊梅)

❀ 做法

1. 楊梅洗淨,完全瀝乾水分之後,先用米酒泡浸6-8小時,再用篩隔起楊梅,瀝乾米酒備用。
2. 取一可密封、寬口及透明的玻璃容器,放入楊梅及桂花陳酒,然後封密,並寫上泡浸日期,放置在陰涼、乾燥及陽光照射不到的地方,讓它靜止擺放約6個月後便可以飲用了。

貼士 ❀

- 可參考第64頁私房楊梅酒

日本水蜜桃酒

日本のモモ酒...234

Japanese peach wine...234

無可否認，日本出產的水果，其賣相、味道、香氣，相比很多其他地方都異常優越！除了巨峰提子之外，水蜜桃也是我的至愛！它艷麗粉紅的外表，甜美多汁的味道，輕易地把我們俘虜了。

🍀 材料

日本水蜜桃　1.6千克(約6個)

米酒　1.8公升

冰糖　600克

🍀 做法

1. 水蜜桃洗淨，抹乾水分，一開為二切開。

2. 取一可封密、寬口及透明的玻璃容器，放入水蜜桃、米酒及冰糖，然後封密，並寫上泡浸日期，放置在陰涼、乾燥及陽光照射不到的地方，偶爾將容器搖晃一下，待冰糖完全溶解之後，便可讓它靜止擺放約六個月後便可以飲用了。

貼士 🍀

選用美國出產的白玉水蜜桃泡浸出來的效果也不錯，其材料、份量也相若。而中國出產的北京水蜜桃也夠甜，夠多汁，但香味就遠遠不及日本水蜜桃了。

提子乾 VODKA
Raisin vodka
レーズンウォッカ

桂圓霖酒
Longan rum
ロンガン・ラム酒

杏脯霖酒
Apricot rum
ドライアプリコット・ラム酒

夏 | Summer

乾果酒

ドライ フルーツ酒...233

Preserved fruit wine...233

浸泡果酒除了新鮮生果之外，還可以選用乾果，但要挑選一些味道及香氣都較濃的乾果，這樣浸泡出來的乾果酒才夠芳香濃郁。由於乾果經過乾燥製成，味道會較新鮮生果酒濃香，因此很適合用來做餐後甜酒，或用以配搭其他甜品，如淋於雪糕、做水果味道的蛋糕等等，塗在一些燒烤的食物上也倍添美味！

今次我挑選了幾款自己喜愛的乾果，配以幾款不同的白酒，浸泡出三款不同口味的乾果酒，味道各有風味，大家亦可隨自己的喜好，將自己喜歡的乾果配以不同白酒，浸泡出適合及屬於自己個性的乾果酒吧！

杏脯霖酒

材料
乾杏脯　八百克
白霖酒　一公升

做法
白霖酒倒入容器內，加入乾杏脯，封密，放在陰涼、乾燥、陽光照射不到的地方，浸泡六個月便可以飲用了。

桂圓霖酒

材料
乾桂圓肉　五百六十克
金霖酒　一公升半（兩樽）

做法
金霖酒倒入容器內，放入乾桂圓肉，封密，放在陰涼、乾燥及陽光照射不到的地方，浸泡六個月便可以飲用了。

提子乾 VODKA

材料
提子乾　五百克
黑加倫子白 VODKA　七百五十毫升（一樽）

做法
VODKA 倒入容器內，放入提子乾，封密，放在陰涼、乾燥、陽光照射不到的地方，浸泡六個月便可以飲用了。

貼士

1. 要買獨立包裝直接食用的乾杏脯及提子乾，會較安全衛生。乾杏脯、提子乾及桂圓肉毋須清洗，否則帶有水分的話，很易變壞。

2. 要買沒有經過硫磺及漂白處理的桂圓肉，這樣浸泡出來的乾果酒才原汁原味，沒有任何添加元素，飲得健康。

黑桑椹霖酒

黑桑の実のラム酒...235
Black mulberry rum...235

桑椹，是養蠶用桑樹的果實，在香港並非是一種常見及普及的生果，它只會在四到六月間當造期短暫出現在本地街市。雖然它外型並不吸引，但它的顏色紫紅艷麗，並有食療作用。中醫認為，桑椹能生津解渴、養髮護肝，並且纖維質豐富，能幫助腸臟蠕動，潤腸通便，其所含的維生素m，更有抗衰老的作用。用它泡成果酒味道濃郁，除了可冰鎮飲用之外，加些鮮奶油煮至濃稠成甜品糕點及雪糕的醬汁都非常合適及美味。

❦ 材料

黑桑椹 1公斤
黑霖酒 1.2公升
冰糖 150克

❦ 做法

1. 黑桑椹洗淨，瀝乾水分後，用廚紙小心將水分吸乾，再放在筲箕內置放1-2小時，以確保水分完全吹乾。

2. 取一可密封、寬口及透明的玻璃容器，放入黑桑椹、黑霖酒及冰糖，然後封密，並寫上泡浸日期，放置在陰涼、乾燥及陽光照射不到的地方，偶爾將容器搖晃一下，待冰糖完全溶解後，便可讓它靜止擺放約三個月後便可以飲用了。

貼士 ❦

1. 要揀選外型飽滿及深紫色的桑椹，因為深紫色才代表桑椹果實成熟，味道較甜及濃；而紅色帶綠的桑椹，味道較酸，並且帶有一種草腥味，泡浸出來的果酒相對地遜色了。

2. 黑桑椹有很多細小縫隙，易藏水分，所以一定要確保水分完全吹乾，因為生水含有很多雜質及微菌，很易令果酒變壞及不耐存。

囍宴私房紅莓酒

10 - 4 - 09

紅莓320g 單晶冰糖

200g 廣西米酒 60ml

自家製ラズベリー酒 ...234
Homemade raspberry wine ...234

♣ 材料

新鮮紅莓　320克
廣西米酒　600毫升
冰糖　200克

♣ 做法

1. 紅莓洗淨，瀝乾水分後用廚紙將水分吸乾，再放在笸箕內置放1-2小時，以確保水分完全吹乾。
2. 取一可密封、寬口及透明的玻璃容器，放入紅莓、米酒及冰糖，然後封密，並寫上泡浸日期，放置在陰涼、乾燥及陽光照射不到的地方，偶爾將容器搖晃一下，待冰糖完全溶解之後，便可讓它靜止擺放約三個月後便可以飲用了。

貼士 ♣

一定要確保紅莓的水分完全吹乾，因為生水含有很多雜質及微菌，很易令果酒變壞及不耐存。

巨峰提子酒

日本巨峰酒...236

Kyoho grape wine...236

用日本巨峰提子醞釀出來的提子酒，高貴冷艷，並散發着陣陣誘人花蜜般的芬芳；這是一道款待貴賓絕佳的餐後甜酒！

🍇 材料

日本巨峰提子　1公斤
三蒸米酒　1公升
冰糖　120克

🍇 做法

1. 巨峰提子洗淨，瀝乾水分後，再用廚紙小心徹底抹乾水分。

2. 取一可密封、寬口及透明的玻璃容器，放入巨峰提子、三蒸米酒及冰糖，然後封密，並寫上泡浸日期，放置在陰涼、乾燥及陽光照射不到的地方，偶爾將容器搖晃一下，待冰糖完全溶解後，便讓它靜止擺放約三個月後便可以飲用了。

貼士 🍇

巨峰提子本身味道較甜，且帶有一種花蜜幽香，所以不宜放太多糖，否則便會蓋過巨峰提子那種特有的尊「貴」風味了。

夏 _ Summer

日本王林蘋果酒

日本の王林りんご酒...236
Japanese Ourin apple wine...236

❧ 材料

日本王林蘋果　1.5千克
(約七個)
三蒸米酒　2.8公升
冰糖　380克

❧ 做法

1. 蘋果洗淨，用廚紙徹底抹乾水分，自然吹乾，用竹籤在蘋果上刺5-6個小洞，讓蘋果的汁液易於滲出。

2. 取一可密封、寬口及透明的玻璃容器，放入蘋果、三蒸米酒及冰糖，然後封密，並寫上泡浸日期，放置在陰涼、乾燥及陽光照射不到的地方，偶爾將容器搖晃一下，待冰糖完全溶解之後，便可以讓它靜止擺放約6個月後便可以飲用了。

貼士 ❧

參考第146頁的自製蘋果果醋

這些水果，全部都可以用來釀果酒。

迷人櫻桃酒

サクランボ酒 ..238
Cherry wine ..238

材料

美國櫻桃　1千克
三蒸米酒　1.2公升
冰糖　500克

做法

1. 櫻桃洗淨，去蒂，瀝乾水分後，再用廚紙徹底抹乾水分。
2. 取一可密封、寬口及透明的玻璃容器，放入櫻桃、三蒸米酒及冰糖，然後封密，並寫上泡浸日期，放置在陰涼、乾燥及陽光照射不到的地方，偶爾將容器搖晃一下，待冰糖完全溶解後，便讓它靜止擺放約三個月後便可以飲用了。

貼士

最好選用美國出產深紫紅色的櫻桃品種，因為它的味道較濃及份外甜美，是泡浸果實酒的上好材料。

私房李子酒

自家製すもも酒 237
Homemade golden plum wine..237

❤ **材料**

黃李　2.2千克
米酒　1.8千克
冰糖　1.2千克

❤ **做法**

1. 黃李洗淨，去蒂，瀝乾水分後再用廚紙徹底抹乾水分，用竹籤在黃李上刺入幾個小洞，讓李子汁液易於滲出。

2. 取一可密封、寬口及透明的玻璃容器，放入李子、米酒及冰糖，然後封密，並寫上泡浸日期，放置在陰涼、乾燥及陽光照射不到的地方，偶爾將容器搖晃一下，待冰糖完全溶解之後，便可以讓它靜止擺放約6個月後便可以飲用了。

貼士 ❤

1. 用李子泡浸出來的果酒酸酸甜甜，很開胃，適合在餐前飲用。有一種類似梅酒的風味，但味道比較溫和，香味亦比較自然。

2. 除了黃李之外，用相近的品種如青布霖、黑布霖、西梅等泡浸出來的效果也相當理想，各有風味。

黑桑棗霖酒

ブラックベリー ラム酒 ...238
Blackberry rum ...238

🍀 材料
新鮮黑桑棗　480克
黑霖酒　550毫升
冰糖　180克

🍀 做法
1. 黑桑棗洗淨，瀝乾水分後用廚紙將水分吸乾，再放在笪箕內置放1-2小時以確保水分完全吹乾。
2. 取一可密封、寬口及透明的玻璃容器，放入黑桑棗、黑霖酒及冰糖，然後封密，並寫上泡浸日期，放置在陰涼、乾燥及陽光照射不到的地方，偶爾將容器搖晃一下，待冰糖完全溶解後，便可讓它靜止擺放約三個月後便可以飲用了。

貼士 🍀
一定要確保黑桑棗的水分完全吹乾，因為生水含有很多雜質及微菌，容易令果酒變壞及不耐存。

荔枝桂花陳酒

ライチ桂花陳酒...239

Aged osmanthus lychee wine...239

這款荔枝桂花陳酒是我最喜歡的果實酒之一，酸甜清香的新鮮荔枝與芬芳出塵的桂花陳酒融合之後，醇厚香甜，就算不喝酒的朋友也會非常喜歡。作為餐後甜酒，或配以味道濃郁的菜式，這絕對是不二之選，配以雪糕更是無懈可擊。新鮮的荔枝仁每年六月左右便開始上市，七至八月更是最當造的季節，大家一定要把握這個黃金時節，為自己泡浸一瓶美味可口的荔枝桂花陳酒！

材料

新鮮糯米糍　1斤
桂花陳酒　1枝
(每枝750毫升)

做法

1. 荔枝洗淨，瀝乾水分，剝殼，去核，起出荔枝肉。
2. 取一可密封、寬口及透明的玻璃容器，放入荔枝肉及桂花陳酒，然後封密，並寫上泡浸日期，放在陰涼、乾燥及陽光照射不到的地方，讓它靜止擺放約三個月左右便可以飲用了。
   (建議泡浸一年或以上效果更佳，我店的荔枝酒都是兩年或以上的。)

貼士

1. 剝荔枝取肉時盡量小心，讓荔枝肉保持完整，不要讓荔枝肉太爛，否則太多果汁直接流出與酒混合的話，會令泡浸出來的荔枝酒非常渾濁；泡浸果酒應讓果實有足夠時間與酒發生關係，讓酒有足夠時間將果實的精華吸取出來。
2. 不同品種荔枝的甜酸度都有所差別，泡浸出來的味道亦會各有不同。以我經驗，用糯米糍泡浸出來的荔枝酒是最香甜可口的，大家亦可因應個人喜歡荔枝酒的濃淡程度，在桂花酒與荔枝的比例上加多或減少。
3. 俗語說「一粒荔枝三把火」，荔枝本身性熱，熱底的人吃多了很容易便會有出暗瘡、牙肉腫痛甚至喉嚨痛等這些反應，只要大家在吃完荔枝後喝適量的淡鹽水，便可以減輕及有助化解荔枝的「熱氣」了。

夏 _ Summer

熱情果菠蘿果醬

熱情果及菠蘿各自都非常吸引，香氣濃烈而甜蜜。用這兩種水果混合煮成的果醬，無論味道、顏色都配合得相當協調及和諧，只要你將果醬的瓶子打開，你便會馬上聞到其濃縮而自然的香味，令人食慾大增。

パッションフルーツとパイナップル ジャム…240
Passionfruit pineapple jam…240

❦ 材料

淨菠蘿果肉　1千克
熱情果肉　300克(約15個)
青檸汁　1個份量
砂糖　300克
鹽　1/2茶匙
水　300毫升

❦ 做法

1. 菠蘿切成細粒，與熱情果肉、青檸汁、砂糖、鹽撈勻，用保鮮紙封好，放入雪櫃，置放一天讓它出水。

2. 第二天將果肉連同排出的水分倒入鍋中，以小火煮滾，撈起浮起的泡沫，盛起涼卻後再放入雪櫃存放一天。

3. 第三天將果肉用篩隔起，將排出的汁液倒入鍋中，加水300毫升，以大火煮滾後改小火煮至汁液開始濃稠變「傑」，然後將先前隔起的果肉回鍋拌勻，以小火繼續煮約五分鐘，此時汁液開始起泡，果肉亦開始變得光亮及已變成果醬的濃稠狀態，這時果醬基本上已經完成。你便可以因應個人喜歡果醬的濃稠程度而選擇熄火，這樣果醬會汁液多些，或再煮長一點時間而濃稠一些。

4. 將果醬盛起，趁熱裝入已經消毒處理的容器內，封好保存，慢慢享用。

貼士 🍀

1. 最好揀選較熟的菠蘿，這樣做出來的果醬口感會鬆軟些，香味亦會濃郁一些。

2. 煮果醬期間會有一些泡沫浮渣浮起，要將它撈起，這樣做出來的果醬便會更晶瑩亮麗，美觀吸引了。

3 煮果醬期間，要不時攪拌果醬，因為糖分較高，以免黏底。

玫瑰水蜜桃果醬

モモのバラ入りジャム...241

Rose peach jam...241

水蜜桃，顧名思義，水潤多汁，甜蜜芳香，用它做成果醬清甜中夾雜着水蜜桃的高雅香氣，再加入玫瑰的淡淡幽香及帶點神秘的紫紅色澤，這果醬定必能成為你夏日早餐的最佳伴侶。

❀ 材料

日本水蜜桃　1公斤(淨肉計，約8個)
砂糖　200克
鹽　1/2茶匙
青檸汁　約1個份量

乾玫瑰　約8粒(剝出花瓣)

❀ 做法

1. 水蜜桃去皮切細粒，與砂糖、鹽、青檸汁撈勻，用保鮮紙包好，放入雪櫃，置放一天讓它出水。

2. 第二天將水蜜桃連同排出的水分倒入鍋中，以小火煮滾，撈出浮起的泡沫，盛起涼卻後再放入雪櫃存放一天。

3. 第三天將水蜜桃及水分倒入鍋中，加入玫瑰花瓣，以大火煮滾後改小火煮至汁液濃稠「傑」身及開始起泡，此時水蜜桃亦開始變得光亮及已變成果醬的濃稠狀態，這時果醬基本上已經完成。你便可以因應個人喜愛果醬的濃稠程度而選擇關火，這樣果醬會汁液多些，或再煮長一點時間而濃稠一些。

4. 將果醬盛起，趁熱裝入已經消毒處理的容器內，可慢慢享用了。

貼士

宜選用日本出產的水蜜桃，而且不要選擇
太生的，這樣做出來的水蜜桃果醬才會香
甜並重，口感細滑，與別不同。

夏 _ Summer

はちみつキウイジャム ..242
Honey kiwi jam ..242

❀ 材料

奇異果　1公斤(約10個)
砂糖　100克
青檸汁　1個份量
鹽　1茶匙

蜂蜜　120克
水　300毫升

❀ 做法

1. 奇異果洗淨，去皮，切成細粒，先與砂糖、青檸汁、鹽撈勻，用保鮮紙封好，放入雪櫃，置放一天讓它出水。

2. 第二天將果肉連同排出的水分倒入鍋中，以小火煮滾，撈起浮起的泡沫，盛起涼卻後再放入雪櫃存放一天。

3. 第三天將果肉用篩隔起，將排出的汁液倒入鍋中，加入蜂蜜及水，以大火煮滾後改以小火煮至汁液開始濃稠變「傑」，然後將先前隔起的果肉回鍋拌勻，以小火繼續煮約五分鐘，至汁液開始起泡，果肉亦開始變得光亮及已變成果醬的濃稠狀態，這時果醬基本上已經完成，你便可以因應個人喜歡果醬的濃稠程度而選擇關火，這樣果醬會汁液多些，或再煮長一點時間而濃稠一些。

4. 將果醬盛起，趁熱裝入已經消毒處理的容器內，封好保存，慢慢享用。

貼士 ❀

1. 可以使用各種不同的蜂蜜如百花蜜、冬蜜、龍眼蜜等，各有風味。

2. 煮果醬期間會有一些泡沫浮渣浮起，要將它撈起，這樣做出來的果醬便會更晶瑩亮澤，美觀吸引，另外在煮的過程中亦要不時攪拌果醬，以免黏底。

雲呢嗱芒果果醬

バニラとマンゴーのジャム...243
Vanilla Mango jam...243

❦ 材料

呂宋芒果　1公斤(淨果肉計)
砂糖　200克
青檸汁　1個份量
鹽　1/2茶匙

雲呢嗱條　2枝

❦ 做法

1. 芒果肉切成細粒，與砂糖、青檸汁、鹽撈勻，用保鮮紙封好，放入雪櫃，置放一天讓它出水。

2. 第二天將芒果連同排出的水分倒入鍋中，以小火煮滾，撈起浮起的泡沫，盛起涼卻後再放入雪櫃存放一天。

3. 第三天將芒果及水分倒入鍋中，將雲呢嗱條剠開，再用刀背小心將雲呢嗱籽刮出，放入芒果拌勻，以大火煮滾後改以小火煮至汁液濃稠「傑」身及開始起泡，此時芒果亦開始變得光亮及已變成果醬的濃稠狀態，這時果醬基本上已經完成。你便可以因應個人喜愛果醬的濃稠程度而選擇關火，這樣果醬汁液會多些，或再煮長一點時間而濃稠一些。

4. 將果醬盛起，趁熱裝入已經消毒處理的容器內，封好保存，慢慢享用。

貼士 ❦

1. 宜選用成熟的芒果，這樣做出來的果醬會更香甜軟滑。

2. 亦可以其他芒果品種來做果醬，風味各具特色。

3. 煮果醬期間會有一些泡沫浮渣浮起，要將它撈起，這樣做出來的果醬便會更晶瑩亮澤，美觀吸引了。另外在煮果醬時亦要不時攪拌果醬，避免黏底。

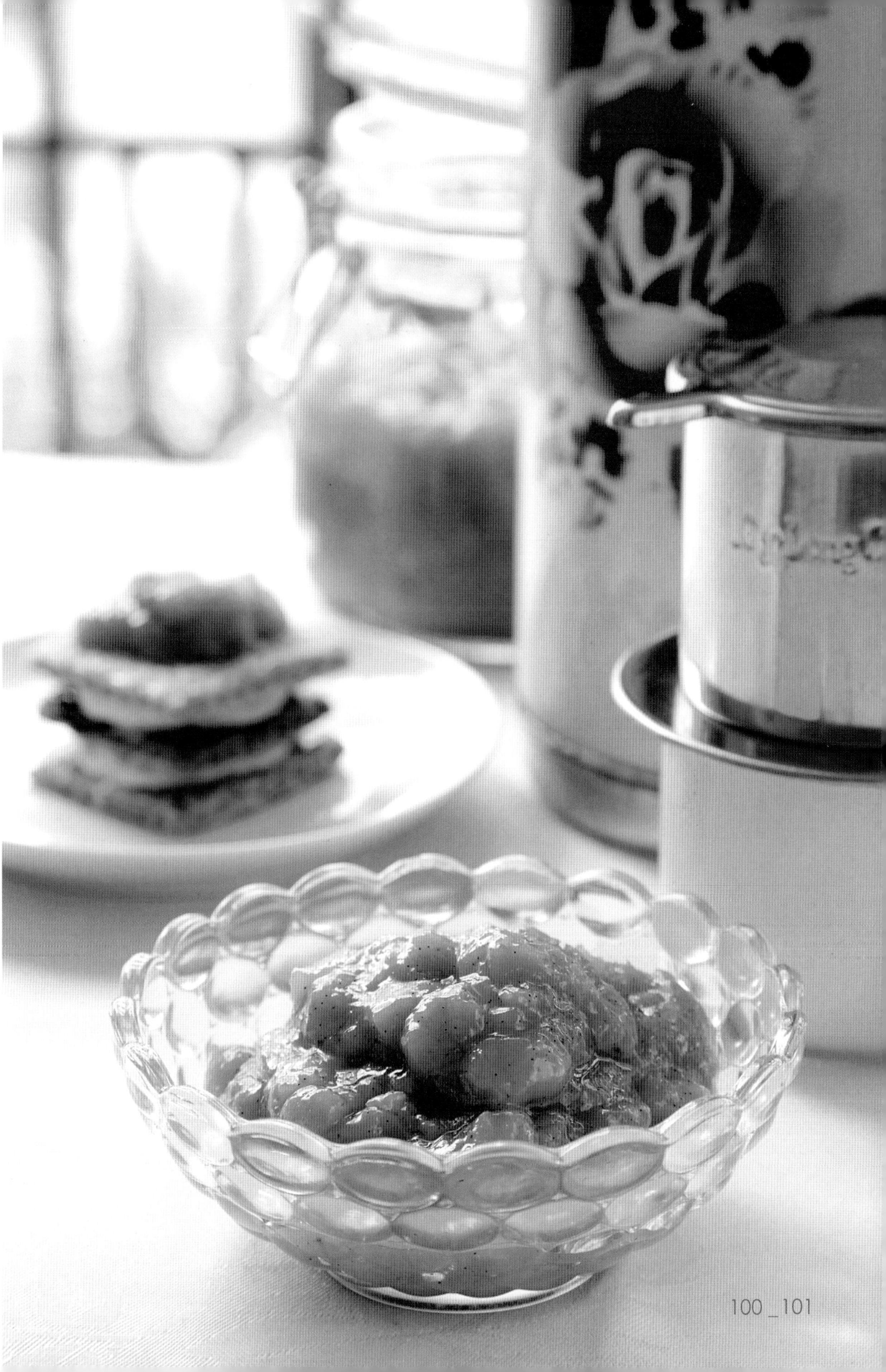

夏 _ Summer

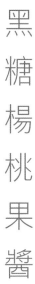

砂糖楊桃果醬

黑糖楊桃果醬

夏 · 果醬

黑糖スターフルーツジャム 244
Star-fruit muscovado jam 244

我很喜歡黑糖那種焦香甘甜的自然風味，很適合用來製造各類果醬，同時亦能帶給傳統果醬不一樣的獨特口味。

❤ 材料

熟楊桃　1公斤
黑砂糖　200克
青檸汁　1個份量
鹽　1/2茶匙

❤ 做法

1. 楊桃去邊，去芯，去籽，再切成細粒，與黑砂糖、青檸汁、鹽撈勻，用保鮮紙封好，放入雪櫃，置放一天讓它出水。

2. 第二天將楊桃連同排出的水分倒入鍋中，以小火煮滾，撈起浮起的泡沫，盛起涼卻後再放入雪櫃存放一天。

3. 第三天將楊桃用篩隔起，將排出的汁液倒入鍋中，大火煮滾後改以小火煮至汁液開始濃稠變「傑」，然後將先前隔起的楊桃回鍋拌勻，以小火繼續煮約五分鐘，至汁液開始起泡，楊桃果肉小開始變得光亮及已變成果醬的濃稠狀態，這時果醬基本上已經完成，你便可以因應個人喜歡果醬的濃稠程度而選擇關火，這樣果醬會汁液多些，或再煮長一點時間而濃稠一些。

4. 將果醬盛起，趁熱裝入已經消毒處理的容器內，封好保存，慢慢享用。

貼士 ❤

1. 你亦可以普通白砂糖代替黑砂糖，做法及分量與黑砂糖一樣(見上文)。

2. 楊桃最好選用較為偏熟一些的，除了果汁較豐富之外，做出來的果醬亦會香濃一些。

3. 煮果醬期間會有一些泡沫浮渣浮起，要將它撈起，這樣做出來的果醬便會更晶瑩亮澤，美觀吸引了。另外在煮果醬時亦要不時攪拌果醬，避免黏底。

桂花陳酒荔枝果醬

醉人芬芳，清甜味美。
桂花陳酒與荔枝是黃金絕配，
除了浸酒之外，
做成果醬亦一定
令你再三回味。

ライチの桂花陳酒入りジャム __245
Aged Osmanthus lychee wine jam __245

🍀 **材料**

新鮮荔枝肉　1公斤(去核淨肉計)
鹽　1茶匙
砂糖　200克
青檸汁　1個份量
桂花陳酒　650毫升

🍀 **做法**

1. 荔枝肉切成細粒，下鹽、砂糖及青檸汁撈勻，用保鮮紙封好，放入雪櫃，置放一晚讓它出水。

2. 第二天將荔枝連同排出的水分倒入鍋中，加入桂花陳酒，以大火煮滾後改以小火煮至汁液開始變成濃稠及起泡，此時荔枝與汁液亦開始變得光亮及已變成果醬的濃稠狀態。這時果醬基本上已經完成，你便可因應個人喜歡果醬的濃稠程度而選擇熄火，這樣果醬會汁液多些及稀些，或再煮長一點時間讓果醬濃稠一些。

3. 將果醬盛起，趁熱裝入已經消毒處理的容器內，封好保存，慢慢享用。

夏 _ Summer

尊貴巨峰提子果醬

プレミアム巨峰ジャム...245

Kyoho grape jam...245

能冠以「尊貴」之名的果醬，我相信只有用日本巨峰提子做成的果醬才有資格。因為它的「造價」確實真的很貴！它帶有花蜜般的香氣，及無可代替的葡萄味道，又的確令人難以抗拒！雖然是貴，但我總覺得，偶爾做一些好東西慰勞一下自己，對自己好一些，是平時辛勞工作的一種回報！

🍇 材料

巨峰提子　1公斤
鹽　1/2茶匙
砂糖　100克
青檸汁　1個份量

🍇 做法

1. 提子洗淨，瀝乾水分，去皮去籽，每粒約分為3-4份，與鹽、砂糖、青檸汁撈勻，用保鮮紙包好，放入雪櫃，置放一天讓它出水。

2. 第二天將提子連同排出的水分倒入鍋中，以大火煮滾後改小火煮至汁液開始變成濃稠及起泡，此時提子與汁液亦開始變得光亮及變成果醬的濃稠狀態。這時果醬基本上已經完成，你便可因應個人喜歡果醬的濃稠程度而選擇熄火，這樣果醬會汁液多些及稀些，或再煮長一點時間讓果醬濃稠一些。

3. 將果醬盛起，趁熱裝入已經消毒處理的容器內，封好保存，慢慢享用。

貼士 🍇

一盒巨峰提子的正常售價約港幣$400-500，頗為昂貴，但用來做果醬的話，你不必要購買最新鮮的。一些大型日式超市通常都會在晚上9：30後便會將擺放太久的生果割價大平賣，有時甚至低至半價……節儉是美德，我很多時都會趁着這個「黃金」時段出沒於各日式超市。

話梅菠蘿果醬

干し梅とパイナップルジャム...246

Dried plum and pineapple jam...246

❧ 材料

淨菠蘿果肉　1公斤
話梅　50克(約10粒)
青檸汁　1個份量
砂糖　250克

水　3杯

❧ 做法

1. 菠蘿肉切成細粒；話梅去核，將肉切碎，與青檸汁、砂糖撈勻，用保鮮紙封好，放入雪櫃，置放一天讓它出水。

2. 第二天將果肉連同排出的水分倒入鍋中，以小火煮滾，撈起浮起的泡沫，盛起涼卻後放入雪櫃存放一天。

3. 第三天將果肉用篩隔起，將排出的水分倒入鍋中，加水3杯以大火煮滾後改小火煮至汁液開始濃稠變「傑」，然後將先前隔起的果肉回鍋拌勻，以小火繼續煮約五分鐘左右至汁液開始起泡，果肉亦開始變得光亮及已變成果醬的濃稠狀態，這時果醬基本上已經完成，你便可以因應個人喜愛果醬的濃稠程度而選擇關火，這樣果醬會汁液多些，或再煮長一點時間而濃稠一些。

4. 將果醬盛起，趁熱裝入已經消毒處理的容器內，封好保存，慢慢享用。

貼士 ❧

可參考第93頁的「熱情果菠蘿果醬」

夏 _ Summer

被遺忘的滋味～阿姨仁稔

此道醃製仁稔，是由我阿姨親自教授，因而取名「阿姨仁稔」。

醃製仁稔是非常傳統的廣東食品，醃製好的仁稔可以當成冷菜、涼果般直接食用，亦可取其肉剁碎加些麵豉、甜酸子薑用來蒸魚、蒸排骨以及蒸雞等等，非常惹味開胃。而用豉油浸泡過仁稔的仁稔汁，你可以加些橄欖油或麻油做成沙律汁，拌以各類涼拌菜式，或加些切碎的辣椒絲用來做沾醬用，更是奇味非常；燜煮肉類，加些仁稔汁來調味，亦會令菜式增添妙不可言的風味。

仁稔的奇特酸澀味道，亦是一道非常可口的醬汁，因為吸收了仁稔除了用來醃製之外，亦可將新鮮仁稔肉加入切碎的五花腩、麵豉、辣椒、菜脯、豆豉炒成仁稔醬，這亦是一道相當經典的廣東家常醬料，非常美味，可惜現在已極少人懂得做了。加上仁稔只在每年初夏約六月短短一個月左右出產，所以更見珍貴，我將仁稔納入此食譜，亦希望仁稔這款獨特而美味的傳統食品可以延續下去，讓更多喜愛烹飪的朋友能得以分享！

阿姨仁稔

おばの仁稔（ヤンニム）...247
My Auntie's pickled Ren Ren...247

因為仁稔的核確實與
人面輪廓有些相似，
因此亦有人將仁稔叫
做「人面」。

🎗 材料

仁稔　5斤
白醋　份量要蓋過仁稔

🎗 醃料

上等生抽　2斤
砂糖　2斤

🎗 做法

1. 將仁稔洗淨後瀝乾水分，用小刀在表皮剝
   十字形，再用菜刀拍鬆整顆仁稔，使其容
   易入味。

2. 將仁稔放入盆內，倒入白醋泡醃約半天(至
   少4小時)(白醋的份量必須要蓋過仁稔)，
   然後用冷開水浸洗，再用箕盛起瀝乾水
   分。

3. 鍋中倒入生抽及砂糖，以慢火煮溶，涼卻
   後備用。

4. 將仁稔放入適當容器內，加入煮好的生抽
   (要蓋過仁稔)，泡醃約一星期便可食用，
   如喜歡吃辣，可以加入指天椒泡醃，更添
   風味。

1. 一定要使用煲沸後冷卻的水來浸洗醃菜,不能用
   生水,因為未經煮熟的水含有很多雜質及微菌,
   容易令醃菜變壞及發霉,使用冷開水便可令醃菜
   更耐存。

2. 在煮生抽及砂糖的時候,一定一定要用慢火,而
   且見到生抽微微開始煮滾時便要關火,因為生抽
   煮溶砂糖的時候,鍋中液態濃度很高,當一滾起
   時便會產生大量的泡沫,若不及時關火,你的廚
   房便將會面臨一塲「浩劫」!

3. 醃製好的仁稔放入密封並經過消毒處理的保存容
   器內,可以存放很長時間(必須放在乾燥、陰涼及
   陽光照射不到的地方),存放越久,味道就越醇
   厚、芳香、美味,我們餐廳現在用的仁稔已經是
   三年前醃製的了。

4. 新鮮仁稔可在較傳統的街市買到,如九龍城、中
   環結志街、灣仔道、跑馬地、北角春秧街街市
   等。

仁稔汁涼拌皮蛋豆腐

仁稔（ヤンニム）汁とピータン豆腐の和え物...248

Tofu and thousand-year egg
cold appetizer dressed in Ren Ren juice...248

❧ 材料

盒裝蒸煮豆腐　1盒
皮蛋　1個
葱花　適量

❧ 涼拌汁

仁稔汁　約1/2杯
麻油　1湯匙

❧ 做法

豆腐切片放在有深度的碟上，皮蛋切細粒鋪仕豆腐上，淋上涼拌汁，灑上適量葱花即成。

貼士 ❧ ━━━━━━━━━━━━━━━━━━━━━━━━━━

1. 仁稔汁使用前宜用密篩將雜質過濾，這樣做出來的汁液會清澈美觀一點。

2. 仁稔汁除了可拌以各式涼拌豆腐外，用來作各式沙律及涼拌醬汁亦非常美味，如喜歡吃辣的話，還可以適量加些辣椒調味，更添風味。

星洲娘惹泡菜

アチャー（シンガポール　ピクルス）...251
Acar (Nyonya assorted pickles)...250

這是一道星洲口味的泡菜，在新加坡很多吃娘惹菜的餐廳及大排檔都可以吃到，非常之有南洋風味；在製作上工序可能會多一些，但當完成後可以品嘗的時候，你便會覺得這些功夫是值得付出的。

🍀 材料 (A)

青瓜　1千克(約兩條)(切條)
椰菜　500克(約半個)(切片)
豆角　150克(約十數條)(切小段)
甘筍　500克(約2條)(切片)
菠蘿　1.5千克(1個)(切粒)
沙葛　500克(約半個)(切片)
中型青、紅辣椒　150克(各約5隻)(切圈)

🍀 材料 (B)

炒香花生碎　適量
炒香白芝麻　適量

🍀 醃料 (A)

鹽　50克
砂糖　1.5千克
白醋　1.5公升

🍀 醃料 (B)

乾葱頭　600克
蒜頭　200克
石栗　約8粒
黃薑　3條(約40克)
**全部舂爛
油　2杯
海南雞飯辣椒醬或蒜茸辣椒醬　約150克

♣ 做法

1. 將切好的材料(A)放入容器中，灑上鹽撈勻，放置一個晚上讓它出水，第二天揸乾排出的水分，備用。

2. 醃料(A)的砂糖及醋略煮，至砂糖完全溶解即關火，待涼備用。燒熱油，先將醃料(B)的乾葱茸、蒜茸、石栗碎及黃薑茸爆香，炒至水分收乾及開始變焦黃，再下辣椒醬炒勻撈起，涼卻後備用。

3. 將醃料全部拌勻，再放入材料(A)撈勻泡醃一天後即可食用，吃時灑上適量炒香花生碎及芝麻伴吃。

貼士 ♣

醃料(B)的材料最好用石舂舂爛，盡量不要用攪拌機，因為攪拌機只能將材料攪碎，而用石舂就可以做到將材料完全舂爛，讓汁液完全滲透出來，讓醃製的食材可以完全吸收，達至最佳的效果。

夏 _ Summer

傳統糖醋醃青木瓜

伝統の青パパイヤ甘酢漬け...249
Traditional sweet and sour green papaya pickles...249

這是在我小學歲月裏，學校小賣部最受歡迎的小食之一。我很清楚記得，那一小塊青木瓜總是螢光橙色的，味道很酸很酸，小賣部會將醃製過的青木瓜雪至冰硬，然後獨立用小膠袋裝着賣，吃完之後，舌頭、嘴唇都是橙色的。在那時候誰也不會理會食用色素是添加物，不健康，但我常覺得，在那個年代，這些用食用色素製造出來的食品，都是那個年代的必然產物，我們的集體回憶！

材料
青木瓜　1個(約1公斤)
鹽　1茶匙

醃料
白醋　500毫升
砂糖　400克

做法
1. 白醋加入砂糖，以小火煮至砂糖溶解後熄火，待涼卻後備用。
2. 青木瓜刨皮、開邊、去籽，再切成片狀，放入容器內，灑上鹽撈勻，放置約半大讓它出水。
3. 將青木瓜排出的水分倒掉，並揸乾水分，放入容器內，注入已涼卻的糖醋，撈勻，泡醃約一至兩天後便可以食用了。

貼士
1. 要揀選不要太生、不能太熟的青木瓜。木瓜切開後，肉呈青白色的，表示味道還未夠，醃製出來只有糖醋味；而太熟的木瓜，因為已經成熟，質地太軟，醃製出來口感太腍，沒有了泡菜應有的爽脆口感。最好選用生熟適中，用手按下去不要太硬，切開後木瓜呈淡黃或淡橙色的最為適合。
2. 醃製過程中，要將青木瓜與醃料至少再撈勻一至兩次，讓青木瓜能均勻入味。

熱情果醃青木瓜

パッションフルーツとパパイヤの漬け物...249
Passionfruit green papaya pickles...249

熱情果醃青木瓜的做法其實跟傳統糖醋醃青木瓜食譜一樣，只是在做法3的時候加入三至四個熱情果肉，與糖醋、青木瓜一起撈勻醃製。加入熱情果後，青木瓜會增添了一份熱情果獨有的濃郁芳香，怡神醒胃！

🍀材料
青木瓜　1個(約1公斤)
熱情果肉　3至4個
鹽　1茶匙

🍀醃料
白醋　500毫升
砂糖　400克

🍀做法
1. 白醋加入砂糖，以小火煮至砂糖溶解後熄火，待涼卻後備用。
2. 青木瓜刨皮、開邊、去籽，再切成片狀，放入容器內，灑上鹽撈勻，放置約半天讓它出水。
3. 將青木瓜排出的水分倒掉，並揸乾水分，青木瓜、熱情果肉放入容器內，注入已涼卻的糖醋，撈勻，泡醃約一至兩天後便可以食用了。

三色醃子薑

我覺得子薑的味道跟愛情的感覺很相似，粉紅嬌嫩，印象浪漫，甜酸中帶着少許辛辣……很貼切吧！踏入五月中，正是子薑最嫩、最多、最當造的時候，今次除了教大家醃製最傳統的糖醋子薑外，還有兩款獨特的創新口味，若你也喜歡子薑的話，相信亦一定會喜歡！

梅子醋醃子薑芽

傳統糖醋醃子薑

醬油醃子薑

傳統糖醋醃子薑

伝統的な新しょうがの甘酢漬け ...253
Traditional sweet vinegar pickled young ginger ...252

❧ 材料

新鮮子薑　2公斤
凍開水　適量(要可浸過子薑)

❧ 醃料

白醋　1公升
砂糖　800克
鹽　3茶匙

❧ 做法

1. 白醋加入砂糖，以小火煮至砂糖溶解後關火，待涼卻後備用。
2. 子薑洗淨，瀝乾水分，切片，放在容器內，灑上鹽撈勻，放置一天讓它出水。
3. 第二天將子薑排出的水分倒掉，再用冷開水浸洗一會，然後將水隔走，並將子薑用手揸乾水分。
4. 最後將已揸乾水分的子薑放入容器內，加入已溶解砂糖的白醋撈勻，泡醃約三天便可以食用了。

貼士 ❧

1. 在步驟3浸洗子薑時，一定要用煲沸後冷卻的水，因為未經煮熟的水含有很多雜質及微菌，容易令醃菜變壞及發霉，使用冷開水可令醃菜更耐存。
2. 子薑味道溫和，辛辣味淡，纖維細緻，口感爽嫩，很適合用來做泡醃菜，而醃製好的子薑更可以配襯各種食材令它變成更美味可口，皮蛋相信是子薑的絕配。除此之外，配以仁稔及麵豉蒸煮各種海鮮及肉類亦相當惹味開胃。
3. 揀選子薑，要選外觀呈象牙色，飽滿，外皮乾淨，聞落有一股薑的淡淡香氣，沒有腐爛以及嫩芽部分呈帶紫的粉紅色便是優質的子薑。
4. 由於子薑水分較多，容易腐爛及不耐存，買回來的子薑要盡快處理，若未能及時處理的子薑，應要抹乾水分，再用報紙或保鮮紙包好放入雪櫃保存。

醬油醃子薑

新しょうがの醤油漬け…253
Soy marinated young ginger…252

🌸 材料

新鮮子薑 2公斤
冷開水 適量(要可浸過子薑)

🌸 醃料 (A)

萬字醬油 500毫升
廣東米酒 200毫升
味醂 200毫升
冷開水 350毫升
砂糖 500克

🌸 醃料 (B)

鹽 3茶匙

🌸 做法

1. 醃料(A)攪勻，以小火煮至砂糖溶解後關火，待涼卻後備用。

2. 子薑洗淨，瀝乾水分，切片，放在容器內，灑上鹽撈勻，放置一天讓它出水。

3. 第二天將子薑排出的水分倒掉，再用冷開水浸洗一會，然後將水隔走，並將子薑用手揸乾水分。

4. 最後將已揸乾水分的子薑放入容器內，加入涼卻後的醃料(A)撈勻，泡醃約三天便可食用了。

梅子醋醃子薑芽

子薑芽是從子薑切下最嫩的發芽部分

葉しょうがの梅酢漬け ...253
Plum-vinegar marinated
young ginger sprouts ...252

🌿 材料

新鮮子薑芽　1公斤
冷開水　適量(要可浸過了薑)

🌿 醃料

梅子醋　500毫升
（梅子醋之做法可參考第37頁
「呷醋有益-私房梅子醋」，
或可於大型日式超市買到）
砂糖　適量
鹽　3茶匙

🌿 做法

1. 梅子醋加入砂糖，調至適合自己的甜度，以小火煮至砂糖溶解後關火，待涼卻後備用。若用現成買回來的梅子醋，則要先試一下梅子醋的甜度才決定是否需要加入砂糖。

2. 子薑洗淨，瀝乾水分，切片，放在容器內，灑上鹽撈勻，放置一天讓它出水。

3. 第二天將子薑排出的水分倒掉，再用冷開水浸洗一會，然後將水隔走，並將子薑用手揸乾水分。

4. 最後將揸乾水分的子薑芽放入容器內，加入已調好的梅子醋撈勻，泡醃約三天便可以食用了。

冬瓜味噌...254
Winter melon fermented soybean paste...254

材料

冬瓜　2公斤
鹽　3茶匙

醃料

廖孖記麵豉醬　350克
砂糖　60克
米酒　1/2杯
甘草　10克

做法

1. 冬瓜刨皮，去籽，再切成約1.5厘米丁粒，放於
   大盤內，加入鹽撈勻，放置一天讓冬瓜出水。

2. 第二天將冬瓜排出的水分倒掉，再用手將冬瓜
   粒的水分揸乾。

3. 醃料撈勻，將已揸乾水分的冬瓜粒放入再撈
   勻，用保鮮紙封好，放入雪櫃醃約三天即成，
   期間要將冬瓜翻動撈勻一至兩次，令其均勻入
   味。

貼士 🐝

1. 由於此醬料主要材料為生冬瓜，水分較多，若放在室溫保存會較容易變壞，
   建議做好後入樽封密，存放於雪櫃會較安全。

2. 麵豉冬瓜醬用途很多，可以用來蒸及燜煮各類海鮮、肉類，非常惹味。

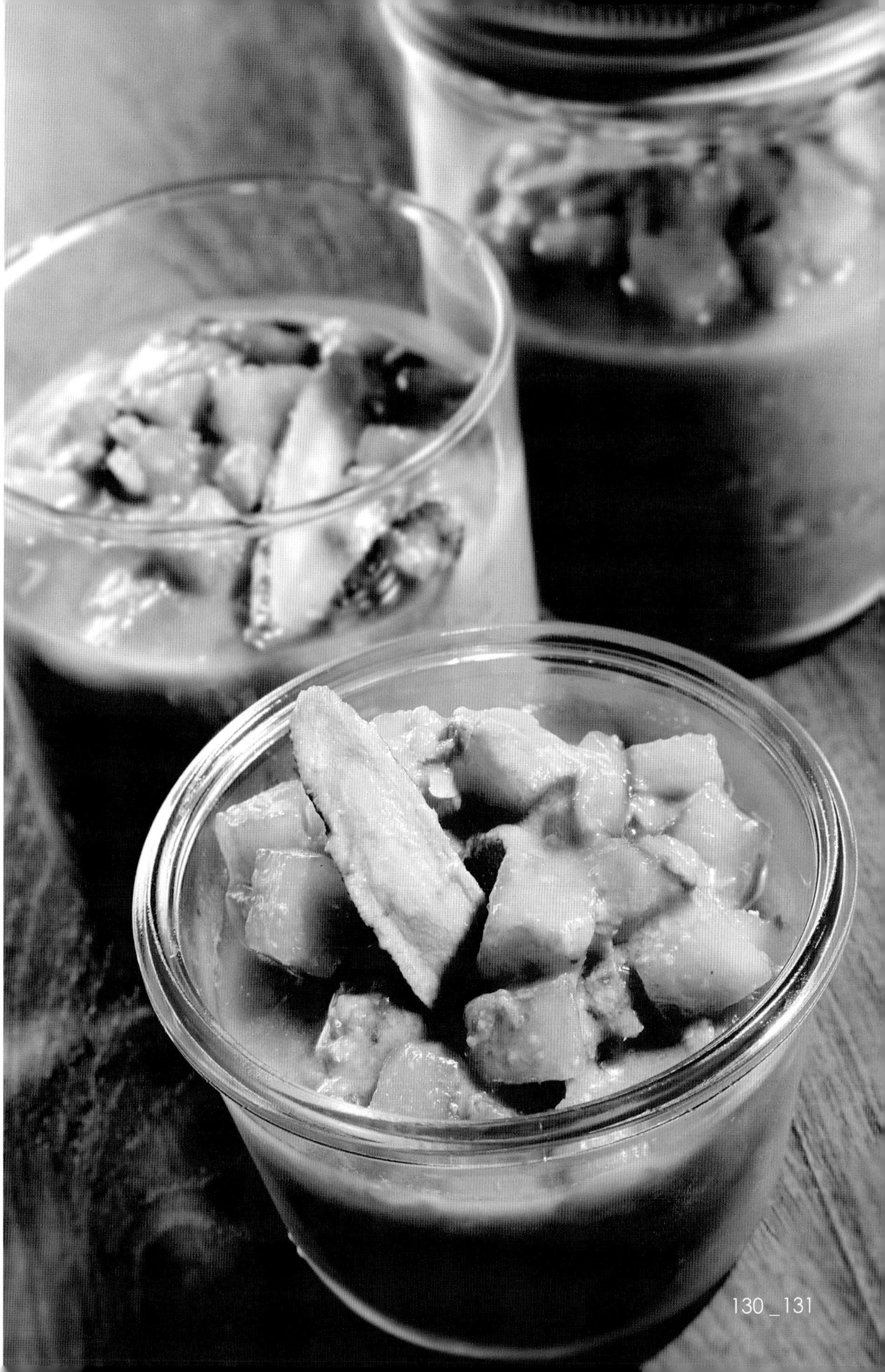

麵豉冬瓜醬蒸鯧魚

冬瓜味噌入りマナガツオ蒸し...254
Steamed pomfret with winter melon
fermented soybean paste...254

🦋**材料**

鯧魚　1條(約600克)
麵豉冬瓜醬　約4-5湯匙
葱絲　適量
油　約2-3湯匙

🦋**做法**

1. 鯧魚劏好洗淨，在魚身底面剠卜十字，使其容易均勻蒸熟及入味。

2. 將麵豉冬瓜醬鋪在魚身，水滾後大火蒸約10-12分鐘，取出放上葱絲，將油燒滾淋在魚身上即成。

貼士🦋

除了鯧魚外，用麵豉冬瓜醬蒸煮各類鹹淡海鮮都非常適合。

麵豉冬瓜醬蒸肉餅

冬瓜味噌入りひき肉蒸し...255

Steamed pork patty with winter melon
fermented soybean paste...255

✿ 材料

絞豬肉　300克
麵豉冬瓜醬　約3湯匙

✿ 醃料

生抽　1湯匙
砂糖　1/2茶匙
雞粉　1/2茶匙
蛋白　1個
生粉　1/2茶匙

✿ 做法

1. 絞豬肉先用醃料撈勻，用筷子同一方向攪拌至豬肉起膠成肉餅狀，再用手拿起肉餅來回「撻」至肉餅富有黏性，再將肉餅放於蒸碟中，醃約1小時。

2. 再在肉餅上鋪麵豉冬瓜醬，水滾後大火蒸約12-15分鐘至熟（視乎肉餅之厚薄），可依個人喜愛灑上適量葱花即可享用。

貼士 ✿

1. 蒸肉餅其實是一道非常家常的小菜，但要蒸出香滑鬆軟的肉餅，其實是有些竅訣。一要選用梅頭肉，因為梅頭肉夠腍，蒸出來的肉餅才夠鬆軟，當然還要用手剁；二這個提議其實有點不太健康，就是一定要加些肥豬肉，這樣蒸出來的肉餅才會滑，閣下在外面吃到那些滑溜肉餅其實也是加了肥豬肉的。

2. 將肉餅放在蒸碟時也不要堆得太厚，因為肉餅太厚的話，相對需要較長的時間才能將肉餅完全蒸熟，這樣肉餅便會過「老」，不好吃了。

タイ風の青マンゴー漬け ...256
Thai pickled green mango ...256

🌿 **材料**

泰國青芒果　3個（約400克）
砂糖　200克
鹽　1茶匙

🌿 **做法**

1. 青芒果洗淨，刨皮。
2. 將芒果肉切成厚片，放於大碗內，倒入砂糖及鹽，撈勻。蓋上保鮮紙，放入雪櫃醃一日。
3. 第二天芒果肉會排出水分及軟化，將芒果肉與汁液撈勻，再放入雪櫃存放一天後即可享用。

貼士 🌸 ━━━━━━━━━━━━━━━━

1. 青芒果可於泰國食品店購買，要選較生及硬身的，醃後才爽口好吃。
2. 醃好的芒果肉存於雪櫃，可保存幾個月，是一道非常開胃怡神的夏日涼果小食。它亦可用來炒牛肉，是一道非常惹味創新的小菜。
3. 而芒果排出的芒果汁液，加些青檸汁及梳打水，便是一杯清新解渴的夏日特飲。

泰式醃青芒炒牛肉

タイ風の青マンゴー漬けと牛肉炒め...257

Stir-fried beef with
Thai pickled green mangoes...257

❧ 材料
醃好之青芒 200克
牛柳肉 200克(切片)
中型青紅椒
各半隻(切片)

❧ 生粉水
生粉 1茶匙
水 適量
＊拌勻

❧ 醃料
鹽 1/2茶匙
雞粉 1茶匙
砂糖 1/2茶匙
生抽 1茶匙
老抽 1/2茶匙
雞蛋 半個
清水 50克

❧ 做法
1. 牛肉先用醃料撈勻醃約半小時，備用。
2. 燒熱油，先將牛肉炒至6成熟，撈起，再起鑊，下青紅椒略炒，牛肉回鑊炒勻，再下青芒炒勻，最後以生粉水埋薄獻炒勻即成。

貼士 ❧
除了牛肉之外，醃青芒炒雞柳、肉片等同樣美味。

青マンゴーのタイ風シーフードサラダ...258
Green mango seafood salad
in Thai style...258

❦ 材料

泰國青芒果　1個
新鮮芫茜　3棵
紅葱頭(乾葱)　1粒
指天椒　1隻
(視乎個人吃辣程度而放多少)
炒香花生　適量
基圍蝦　約8隻
小型魷魚　半隻

❦ 沙律汁

青檸汁　5湯匙
魚露　2 1/2湯匙
砂糖　3 1/2湯匙

❦ 做法

1. 材料全部洗淨，青芒果刨皮切片，芫茜切段，紅葱頭、指天椒切圈，花生壓碎，基圍蝦灼熟去殼，魷魚灼熟切圈，沙律汁撈勻備用。

2. 將所有材料(除花生碎外)放入沙律盤中，加入沙律汁撈勻上碟，最後在沙律面灑上花生碎即可享用。

貼士 ❦

1. 泰國青芒果可於灣仔道石水渠街市、九龍城等一些泰國雜貨店買到。

2. 花生碎一定要最後才下，太早下的話花生碎便會吸收了沙律的水分，便不夠香脆，不好吃了。

沖繩黑糖醃白玉涼瓜

用糖來醃漬泡菜，可以去除瓜菜的菜腥味，適合用來泡醃味道較強烈，容易變壞及帶苦澀味的瓜菜，同時還能帶出瓜菜水分中的酵素及微生物，讓泡菜更爽口甘甜。

白ゴーヤの沖繩黑砂糖漬け ...259
White bitter melon marinated in Okinawa Kurozatou...259

❧ 材料
白玉涼瓜　2個(約850克)
鹽　3茶匙

❧ 醃料
沖繩黑糖　150克
黑醋　20克
薑汁　30克

❧ 做法
1. 涼瓜洗淨，開邊，去籽，再切成片狀，並切去涼瓜的棉質內膜，放入容器，灑上鹽，撈勻，放置一天。
2. 第二天將涼瓜飛水，並馬上浸冰水，待完全涼卻後撈起，並揸乾水分。
3. 將黑糖舂碎，與黑醋及薑汁拌勻，再放入涼瓜撈勻，泡醃約二至三天便可食用了。

貼士 ❧
1. 涼瓜的苦味主要來自瓜瓤的棉質內膜，你可以將此部分切去，但若你同我一樣都鍾情於涼瓜那種特有的甘苦味道，你可以將它保留，況且，這亦是涼瓜營養最豐富的地方。
2. 涼瓜是較易變壞及不耐存的蔬菜，所以在做法步驟(2)時一定要徹底揸乾水分，並放在雪櫃保存，這樣便萬無一失了。
3. 很多人以為白玉涼瓜是新的涼瓜品種，其實它與傳統同一形狀的綠色涼瓜是同一品種，只是在種植的時候，農夫將涼瓜用黑布袋套着，讓涼瓜不能接觸到陽光，不能達至光合作用，於是便不能產生綠色素，種出來就是現在我們見到的白玉涼瓜了。

秋 Autumn

自製
蘋果果醋

以下兩款蘋果果醋，製作份量、蘋果
品種雖不同，但味道同樣香濃芬芳，
可口怡人。

王林リンゴ酢 [261]
Japanese Ourin apple Vinegar [261]

日本王林蘋果醋

❦ 材料

＊小份量製作
日本王林蘋果　2.3公斤(約10個)

❦ 醃料

白醋　2.5公升
麥芽糖　500克
冰糖　800克

日本富士蘋果醋

富士リンゴ酢 [261]
Japanese Fuji apple vinegar [261]

❦ 材料

＊大份量製作
日本富士蘋果　4.6公斤(約20個)

❦ 醃料

白醋　5公升
麥芽糖　1千克
冰糖　1.6千克

🍀 做法

1. 蘋果洗淨，用廚紙徹底抹乾水分，並讓它放置數小時以確保殘餘水分完全揮發，再用竹籤在蘋果上刺些小孔，使汁液容易滲出。

2. 將容器完成消毒過程後，放入蘋果，加入白醋、麥芽糖及冰糖，然後封好，放在陰涼、乾燥及陽光照射不到的地方，偶爾將容器搖晃一下，泡醃最少六個月後至蘋果已變成皺皮及體積縮小便可以使用了。

貼士 🍀

1. 要揀選蘋果味較濃的蘋果品種，這樣泡醃出來的蘋果醋便更香濃芬芳，可口怡人了。日本品種的王林蘋果及富士蘋果是我嘗試過最好的選擇，可於大型日式超市或九龍城的著名生果店均可買到。

2. 一定要確保蘋果表面的水分完全吹乾揮發，這樣蘋果醋便能更長時間保存，若泡醃過程中存有生水，蘋果醋便很易變壞。

ミニトマトのリンゴ酢漬け...262
Pickled cherry tomatoes in apple vinegar...262

這道健康小食做法極為簡易，你可以使用自己泡浸的蘋果醋，亦可以在市面上購買一些現成的果醋泡製。果醋與車喱茄互相吸收了大家的味道之後，香味更豐富，味道亦更有層次，泡醃完成後的車喱茄可以是一道賣相精緻的涼菜，亦可以是一道日常健康小食。習慣在辦公室吃飯盒的朋友，可以預先做好帶回公司，在匆忙的午餐補充一下平時攝取不足的蔬菜纖維。而吸收了車喱茄味道的果醋，更是一杯怡神有益的健康飲品；你還可以運用自己的創意，將它們變成更多不同的組合及吃法：做薄餅、拌沙律，甚至將它做成雪葩等，效果都相當不錯。

🍀材料
蘋果醋　1公升
車喱茄　約1公斤

🍀做法
1. 車喱茄洗淨，先用小刀於車喱茄底部剕一個小十字。煲滾水，將車喱茄放入滾水中灼約十數秒，至車喱茄表皮微微爆開，將車喱茄馬上撈起浸入冰水，再用小刀小心將車喱茄皮剝掉，瀝乾水分備用。
2. 取一容器，放入車喱茄，倒入蘋果醋(蘋果醋要蓋過車喱茄)，封好，泡醃一天左右便可以食用了。

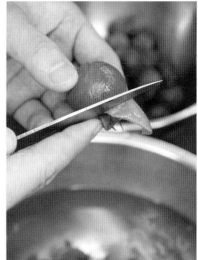

貼士 🦋

1. 用滾水灼車喱茄的時間切勿太長，只需十數秒，至表皮微微爆開便要馬上撈起，否則車喱茄過熟，口感便會太腍，失去生番茄原有的清爽口感；而泡醃好的果醋及車喱茄放入雪櫃內，可以保存三至四星期。

2. 除了使用蘋果醋之外，你還可以選用一些其他口味的果醋代替，如檸檬醋、梨醋、葡萄醋等等，風味各有特色。

車喱茄果醋啫喱

ミニトマト酢のゼリー...262

Cherry tomato vinegar jelly...262

⚘ 材料

泡醃過車喱茄的果醋　1公升
魚膠片　50克(約10片)

⚘ 做法

1. 魚膠片先用冰水浸約10分鐘使其軟身，然後揸乾水分備用。

2. 果醋加熱至攝氏60-70度左右，放入魚膠片邊煮邊攪拌至完全溶解，再用密篩過濾以去除雜質及未溶解的魚膠片殘渣。

3. 將果醋倒入預備好的容器或盤中，涼後放入雪櫃冷凍凝固成啫喱狀即成。

貼士 ⚘

1. 果醋啫喱可直接食用，亦可切成自己喜愛的形狀配以各類食品或飲品，各俱風味。

2. 可隨自己喜歡啫喱的軟硬度而加減魚膠片的份量。

車喱茄果醋特飲

ミニトマト酢のドリンク...263
Cherry tomato vinegar drink...263

🍀做法

　　將泡醃過車喱茄的果醋冰凍後倒入小杯中，再放入
　　一粒泡醃好的車喱茄即可。

低溫松露油風乾番茄

ドライトマトのトリュフオイル 入り…264
Low-temperature air-dried tomatoes with truffle oil…264

🌿 材料

中型番茄　約十數個
(羅馬茄或蛋茄)

🌿 調味料

松露油、鹽、胡椒粉　各適量

🌿 做法

1. 番茄洗淨，瀝乾水分，開邊切開。預熱焗爐60℃，在焗盤上鋪上錫紙，放上番茄，再灑上少許鹽及胡椒粉(如買回來的番茄太酸的話，可以加入少許砂糖)，放入焗爐風乾約6小時，至番茄呈半乾濕狀態，淋上適量松露油，再焗約半小時左右即成。

2. 從焗爐取出番茄，涼卻後便可放入食物盒保存，慢慢享用。

貼士 🌿

1. 風乾期間，焗爐的門要打開少許，讓蒸氣可以散發，以免蒸氣困在焗爐內，令番茄不夠乾身。

2. 松露油要最後才下，否則油分會將水分鎖着，需要較長時間風乾；而松露油亦因長時間焗烤而蒸發，香氣流失。

3. 做好後的風乾番茄，可以直接食用，做各式沙律配料及各類涼拌菜式，而風乾後的番加皮亦較韌，難以消化，建議吃時可將番茄皮剝掉。

4. 如家中有食物風乾機，亦可以用相同做法炮製。

秋_Autumn

冰鎮薄荷梅子番茄

冷やしミント、梅、トマト ...265

*Cold tomato appetizer in plum dressing
with mint leaves* ...265

♣ 材料
原個番茄 5-6個
新鮮薄荷葉 適量

♣ 醃料 (A)
話梅 40克(約十數粒)
碎冰糖 40克
青檸檬汁 1湯匙(約1個份量)
清水 1杯

♣ 醃料 (B)
意大利黑醋 1湯匙
橄欖油 適量
梅子粉 約2茶匙

♣ 做法
1. 將醃汁(A)煮滾,涼卻後加入醃汁(B)的黑醋及橄欖油攪勻,放入雪櫃冷藏備用。
2. 番茄放入滾水灼約十數秒,撈起馬上浸冰水,至表皮爆開,剝去番茄皮,放入雪櫃冷藏至冰凍,備用。
3. 食用時,將番茄橫切成片,於每片番茄之間鑲入薄荷葉,淋上冰凍醃汁及灑上適量梅子粉伴吃。

貼士 ♣
1. 另一種吃法是,將番茄切片與薄荷葉放入調好的醃汁內,放於雪櫃泡醃約2-3小時至入味後上碟,灑上梅子粉享用,絕對是夏日佳品。
2. 宜選用外國品種的沙律用番茄,因為可以直接生吃,沒有本地番茄生吃時的菜腥味。

私房
五香鹹蛋

~「保證零失敗」
簡易自家製作~

看電視新聞，經常看到一些駭人聽聞的食物產品，其實是用一些匪夷所思的方法「製造」出來，像近期大陸有十份一的食肆，都是用一些從坑渠收集回來的坑渠油，再提煉還原賣給食店，簡直缺德兼恐佈。雖然眼不見為乾淨！但這些報道，總會讓人對大陸某些食品產生陰影，尤其是一些醃製類食物。

因此，我很鼓勵大家自己製作食物，一來這是生活上的一種樂趣，二來乾淨衛生，吃得安心！

這個醃鹹蛋，講明「零失敗」，的確非常簡單易做，保證成功！

用傳統方法醃製鹹蛋一般是將禾稈草燒成灰，再混合鹽及水開成漿包裹着鴨蛋醃製約一個月而成！

而今次教大家的私房五香鹹蛋，加入了不同的香料，令香味更富層次，絕對是私房獨家風味。

自家製香辛料と塩漬けの卵 ...266
Homemade Foolproof Five-spice salted eggs ...266

❦ 材料

鴨蛋、雞蛋　各10個
水　2公升
粗鹽　500克
花椒、八角、草果、
丁香、桂皮　各 10克

❦ 做法

1. 鴨蛋、雞蛋洗淨，抹乾水分。
2. 水、鹽及香料一起煮滾，使香料出味，攪拌均勻使
   粗鹽完全溶解，待涼備用。
3. 待水分完全冷卻後，放入鴨蛋及雞蛋，用保鮮紙封
   好，靜止擺放醃約一個月後即成私房五香鹹蛋了！

成功啦！就係咁簡單，
只是需要一點耐性的等待而已。

WOW！晶瑩剔透的鹹蛋黃，
比起在街市買的，一點也不遜色！

貼士 🌸

1. 由於水分含鹽的濃度較高，因此浮力很強，放入蛋之後，蛋會全部浮起，浮出水面的蛋便未能完全吸收味道，所以，你應把鹽水注至容器剛滿，然後蓋上砧板，再壓上重物，這樣所有蛋便可完全浸在鹽水中，令味道均勻吸收了。若不能這樣做的話，你便要每隔一星期將蛋輕輕攪拌一會，讓它均勻吸收味道。

2. 加入香料是一個新嘗試，這樣醃製出來的鹹蛋便有一種與別不同的私房香味了。

3. 用這種方法醃製出來的鹹蛋無論外觀及味道都跟傳統鹹蛋沒甚麼分別，蛋黃同樣有「起沙」的效果，今次我還用了很少人用來做鹹蛋的雞蛋來醃製，效果也相當不錯，除了蛋黃一樣「起沙」之外，蛋黃也有一層香氣四溢的蛋黃油，相當吸引，只是雞蛋的蛋黃沒有鴨蛋的蛋黃大。

4. 鹹蛋醃夠一個月便可以撈起食用或放入雪櫃存放，不要繼續泡在鹽水裏，否則越泡越鹹。

五香鹹蛋炒涼瓜

香辛料塩漬け卵とゴーヤ炒め...267
Stir-fried bitter melon with
Five-spice salted eggs...267

♣ 材料

涼瓜　2條(約400克)
自製五香鹹蛋　2個
蒜頭　約3粒(拍鬆)
紅椒　半隻

♣ 調味料

鹽　1/3茶匙
砂糖　1/2茶匙
雞粉　1/3茶匙

♣ 做法

1. 涼瓜洗淨，去籽，切片，飛水後瀝乾水分。鹹蛋蒸約15分鐘至熟切粒，蒜頭拍鬆，紅椒切片備用。
2. 鑊中燒熱油，先爆香蒜頭，下涼瓜、紅椒炒勻，下鹹蛋粒及調味料炒勻後即可上碟享用。

貼士 ♣

涼瓜先用約1/2茶匙鹽撈勻醃約半小時，讓它出水後揸乾才飛水，這樣除了可減低苦味之外，還會令炒出來的涼瓜更爽口。

秋 _ Autumn

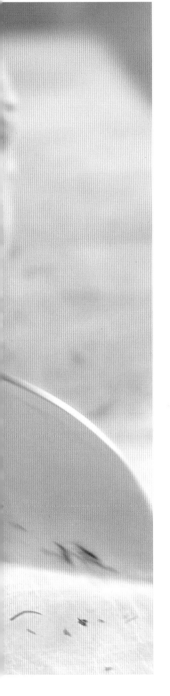

清酒香草醬油漬三文魚

サーモンの清酒、ハーブ、醤油漬け...268
Salmon Shoyuzuke with sake and dill...268

在繁忙緊張的都市裏生活，偶爾停下來，做一些自己喜愛的料理，實在是非常鬆弛神經的調劑。有時候，料理不一定需要好「大陣仗」及很複雜的，一些簡易而又效果出色的食譜會為生活帶來無限樂趣。今次跟大家分享的這道菜式，容易得來更是味道鮮美又有特色，而且一學就會，吃得開心！

🍀 **材料**

新鮮三文魚　一塊（約500克）
刁草　4-5棵

🍀 **醃料**

日本清酒　250毫升
味醂　250毫升
萬字醬油　250毫升
蜂蜜　約1 1/2湯匙

🍀 **做法**

1. 先將清酒與味醂倒入鍋中燒熱，至開始煮滾時小心在上面用火點着，至火焰熄滅後關火，涼卻後倒入萬字醬油及蜂蜜拌匀，然後放入雪櫃雪凍。

2. 待醃料完全冷凍之後從雪櫃取出，將三文魚放入，用保鮮紙蓋好，放回雪櫃泡醃一天。

3. 第二天將三文魚取出，瀝乾汁液，放在保鮮紙上，將刁草切碎，鋪在三文魚上，用保鮮紙包好，再醃一天後即可享用。

貼士 🌿

1. 醃料一定要雪凍才可放入三文魚，因為新鮮三文魚本身一定要冷凍保存，若醃料溫度高於三文魚的話，三文魚便會滋生細菌，容易變壞。

2. 我試過用紹興酒及玫瑰露代替日本清酒來做此菜，效果也相當不錯，各俱風味。

3. 刁草又稱蒔蘿，是一種非常普遍應用的香草，它具有茴香的香氣。新鮮蒔蘿是魚類及海鮮的最佳拍檔，蒔蘿羽毛狀的葉子很適合用在醃漬的食物上，除此之外，亦可剁碎做成各式醬料，亦可加入湯中及配以各類沙律，增添風味。

胡椒海鹽醃三文魚

サーモンのコショウ塩漬け ..269
Pepper-scented salted salmon...269

❧ **材料**

新鮮三文魚　一塊(約500克)

❧ **醃料**

日本清酒　2湯匙
胡椒粒　2湯匙
上等海鹽　1湯匙

❧ **做法**

1. 三文魚先用廚紙抹乾水分，然後均勻抹上清酒，放於雪櫃醃約2小時。

2. 醃魚期間將胡椒粒用石舂舂碎，與海鹽以小火炒至開始焦黃及散出香氣，撈起涼卻後均勻搓抹於三文魚的上面，再用保鮮紙包好，放入雪櫃醃一天，第二天拆開保鮮紙，再切成適當大小，便可以享用了。

貼士 ❧

1. 胡椒及鹽只塗抹於三文魚其中一面便已足夠。因為其一若將醃料塗抹在整塊三文魚的話，味道會過鹹；其二鹽分與食材溶合之後會令食材出水，這樣會令完成品太濕，影響味道之餘賣相亦欠佳。

2. 整個醃製過程須存放於雪櫃，以確保三文魚之新鮮及避免變壞，你亦可以先試做一次，試味後才因應自己喜歡的鹹淡口味而決定鹽的分量。

秋 _ Autumn

鶏のコショウ入り塩漬け ...270

Pepper-scented salted chicken ...270

雞,相信沒有人不喜歡吃!而且,世界各地都有各式各樣數之不盡的烹調方法,無論你用那一種方式去做:蒸、煮、烤、焗、炸、燒、燜、燉,甚至煲湯等等,我都覺得,雞都是一種最容易達到預期效果,易於處理而且非常美味的食材之一。

這道胡椒鹹雞,不單止聞起來香,吃起來惹味醒胃,還可以一次過多隻製作,就像臘腸、臘肉、臘鴨一樣,做好後放於雪櫃長時間保存,慢慢享用。

我經常強調,自己做,衛生安全,吃得更安心!

♣ 材料

雞 2隻

♣ 醃料

胡椒粒 5湯匙
海鹽 5湯匙

♣ 做法

1. 雞洗淨,斬去頭、尾,中間劏開一分為二,用廚紙或布徹底吸乾水分,備用。

2. 胡椒舂碎,與海鹽以慢火炒至開始焦黃及散出香氣,倒出涼卻後,以適當分量均勻擦滿雞身內外,(雞腔內部朝上)置於可疏氣及有底盤的筲箕內,然後放於雪櫃(保鮮冷凍層便可,不能放於雪藏格)一星期。

3. 一星期後,雞隻漸漸收乾水分,會開始變乾及變得較硬,將雞反轉,放回雪櫃,再放置一星期,直至整隻雞完全乾透,接着便可用食物保鮮袋遂件包好封密,放在雪藏格繼續保存。

4. 食用時將雞從雪櫃取出,以室溫自然解凍後,便可以各種烹調方式煮熟享用。

🌸 建議食法

a. 將雞略為沖洗以去多餘胡椒碎及鹽分；水滾後大火蒸約18-20分鐘至熟，涼卻後斬件食用，或撕成雞絲伴以新鮮芫茜、葱絲，淋上麻油，灑上炒香芝麻拌吃。

b. 炮製成胡椒鹹雞煲仔飯。

c. 配些薑片、葱段煲鹹雞粥……等等，各適其適。

貼士 🌸

1. 雞身在擦上胡椒碎及鹽之前，一定要徹底抹乾水分，否則太濕的話需要更長時間風乾及容易變壞。

2. 炒胡椒碎及海鹽時，一定要用慢火，這樣鹽便有足夠時間去吸收胡椒的味道及香氣，做出來的胡椒鹹雞便更美味了。

星洲醃青辣椒

シンガポール風の青唐辛子ピクルス...272

Singaporean pickled green chillies...272

若你去過新加坡的話，在很多大排檔、麵檔，甚至大型餐廳，侍應生總會在枱上放上一碟醃製的酸辣椒。我很喜歡這一小碟醃菜，開胃醒神，加些醬油做湯渣的沾醬更是美味。此酸辣椒在香港的餐廳較少見到，你不妨一次過做多些，隨時加在喜歡的食物上，慢慢享用。

🌸 材料

中型青辣椒　2公斤

🌸 醃料

白醋　1公升
砂糖　800克
鹽　3茶匙

🌸 做法

1. 白醋加入砂糖，以小火煮至砂糖溶解後關火，待涼卻後備用。

2. 青辣椒洗淨，瀝乾水分，切圈，放入容器內，灑上鹽撈勻，放置一天讓它出水。

3. 第二天將青辣椒排出的水分倒掉，再用手揸乾水分放入容器內，加入已溶解砂糖的白醋撈勻，泡醃約兩天便可以食用了。

貼士 🌸

1. 無論粥、粉、麵、飯或其他餸菜，這款酸酸辣辣的醃青辣椒，都會令你增添食慾，為食物帶來更出色的風味。

2. 你亦可以將此醃青辣椒混合剁碎的豆豉，用來蒸煮海鮮及肉類都相當美味。

3. 青辣椒一定要揸乾水分，否則有生水的話，泡醃出來的醃菜便不耐存，容易變壞及發霉。

秋
•
辣
椒

秋 Autumn

私房剁椒醬

自家製の唐辛子ソース...271
Homemade chopped chilli sauce...271

❤ 材料

中型紅辣椒　1公斤
老薑　50克(洗淨，切片)
蒜頭　80克(洗淨，切片)
豆豉　50克(洗淨，瀝乾
水分略為切碎)
油　300毫升

❤ 調味料(A)

白醋　100毫升
粗鹽　40克
雞粉　5克(約1/2湯匙)
雞粉　10克(約1湯匙)
辣椒油　90克
麻油　60克
魚露　100毫升

❤ 調味料(B)

生蒜片　適量
啤酒　約30克

❤ 做法

1. 辣椒洗淨，瀝乾水分，去蒂，開
　 邊，去籽，切碎或用攪拌機攪碎。
2. 辣椒碎用白醋、粗鹽、雞粉5克撈
　 勻，醃一天後將排出的水分搾乾，
　 (水不要)。
3. 鑊下油燒熱，先下薑片，以小火爆
　 香至開始變黃，下蒜片爆香至開始
　 變黃，再下豆豉，繼續以小火炒至
　 出味，下辣椒，改大火炒煮至辣椒
　 變軟及辣椒皮開始捲起，下辣椒
　 油、麻油、魚露、雞粉10克撈勻煮
　 滾後便可關火，倒出待涼卻後便成
　 美味萬用私房剁椒醬。

貼士 ❤

1. 做好的剁椒醬最好用來烹調蒸類的菜式，尤其是海鮮，相當惹味，如蒸大魚頭、魚類、
　 貝殼類都適合。
2. 使用時，每150克之剁椒醬加入啤酒約30克及適量生蒜片撈勻，鋪在所烹調的材料上同
　 蒸，因為加入啤酒可令烹煮的食材有提鮮及令肉質有更嫩滑的作用，尤其是牛肉。
3. 若不喜歡太辣的話，可適量減少辣椒油的份量。

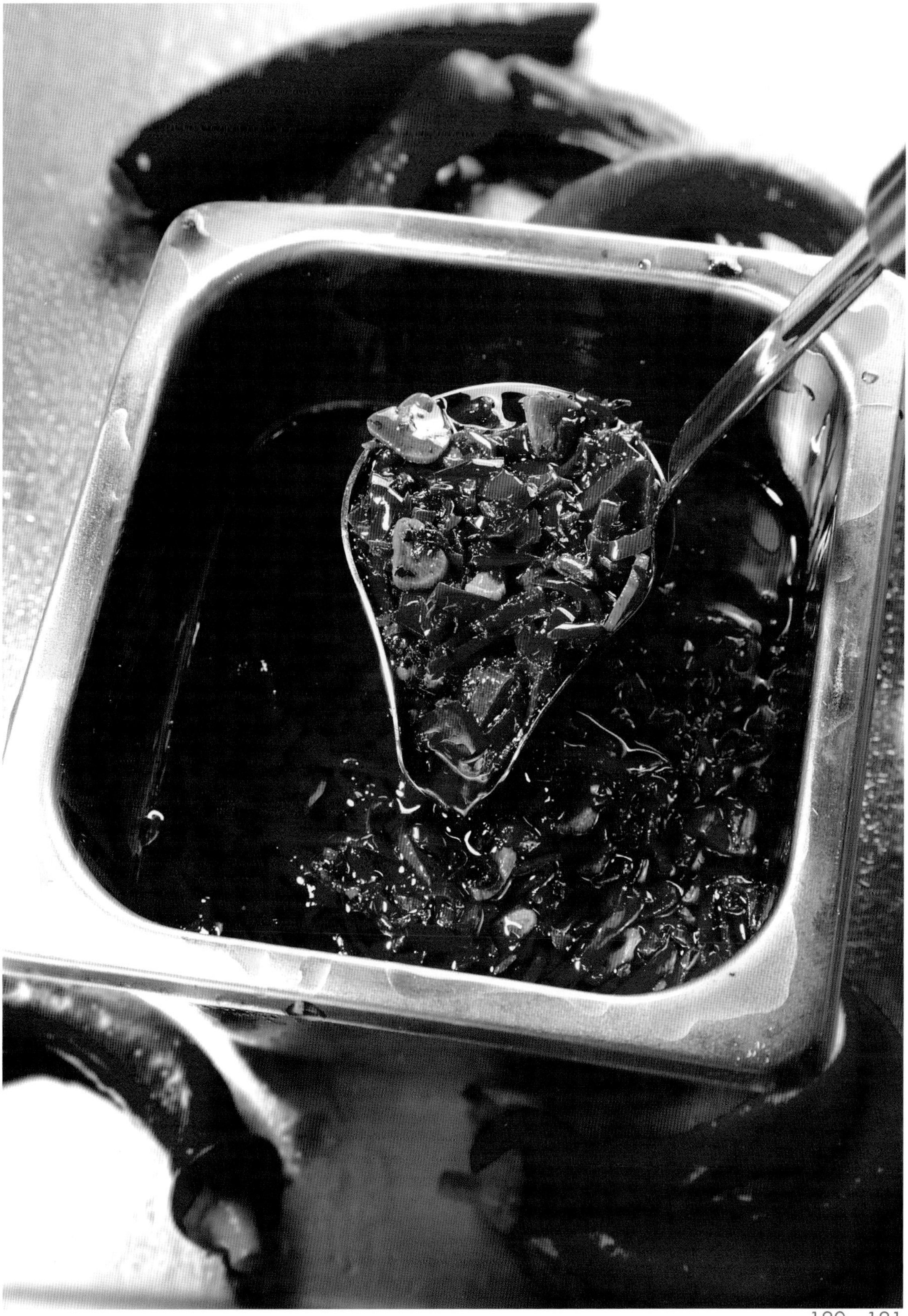

秋 _ Autumn

私房剁椒醬蒸大魚頭

自家製唐辛子ソース入りコクレンの頭蒸し ..272
Steamed fish head with
homemade chopped chilli sauce...272

🍀 **材料**

大魚頭 1個（約900克）
日本葛絲 適量
葱花 適量

🍀 **調味料**

私房剁椒醬 約250克
啤酒 約50克
蒜頭 2-3粒（切片）

🍀 **做法**

1. 葛絲浸軟，盛起，放在深碟內。
2. 大魚頭去腮，洗淨，瀝乾水分；放在葛絲上，將調味料撈勻，均勻鋪在魚頭上，水滾後以大火蒸約12-15分鐘，取出灒上適量葱花即可享用。

貼士 🍀 ━━━━━━━━━━━━━━━━━━━━━

1. 蒸魚頭的時間可因應魚頭的大小而調整，而剁椒醬的份量亦可因應個人喜好吃辣的程度而加多或減少。
2. 剁椒醬同時亦適合用來蒸煮其他鹹淡水魚類及海鮮。
3. 葛絲與魚頭同蒸，用以吸收魚頭的鮮味及剁椒醬的味道，你可以粉絲、薯麵之類代替，各俱風味！

美極黑醋醃茄子

ナスのマギーと黒酢漬け...263
Eggplant pickled with Maggi's seasoning and balsamic vinegar...263

🍀 材料
日本茄子　兩條

🍀 醃料
美極鮮醬油　150克
意大利黑醋　50克
砂糖　60克
麻油　60毫升

🍀 做法
1. 茄子洗淨，水滾後蒸約15分鐘至茄子軟身及熟透，盛起涼卻後開邊備用。
2. 將醃料調勻，放入茄子泡醃一晚即可食用，期間要將茄子小心翻動，使其可均勻入味，吃時將茄子上碟，再淋上適量醃汁伴吃。

貼士 🍀
1. 要揀選圓潤厚肉的茄子，這樣吃起來的口感便會豐厚飽滿一些。
2. 享用前，你亦可在茄子上加添一些自己喜歡的配料如肉鬆、炸乾葱、紫菜絲、炒香芝麻之類，更添風味。

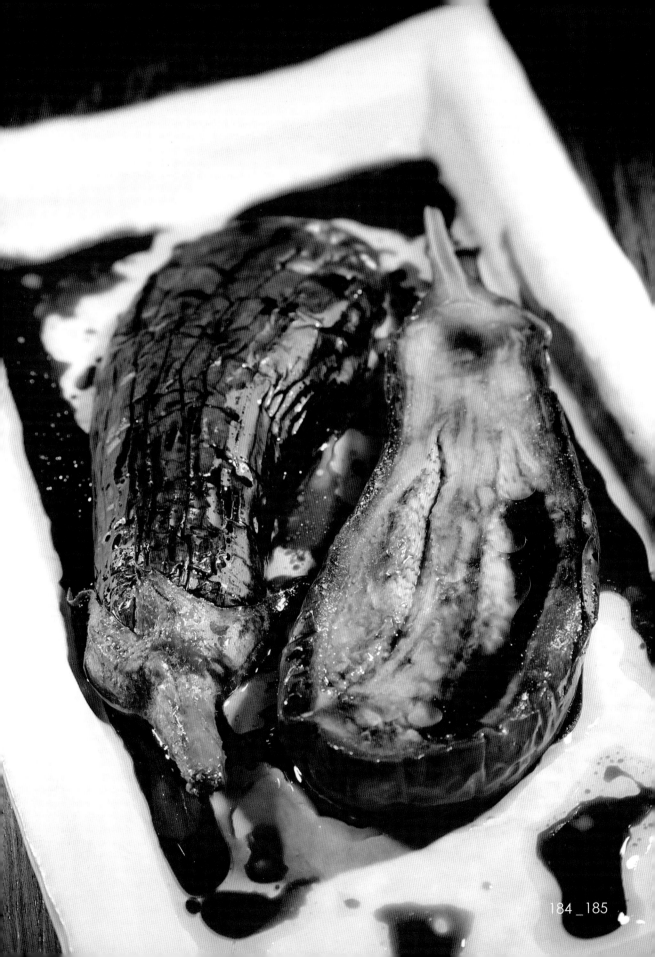

冬
winter

私房陳皮

小時候住在佛山，還記得在那個年代，到時候就會有些「收買佬」打著銅鑼推著車子在街上經過，叫喊著甚麼收買爛銅爛鐵之類的口號，於是各家各戶就會在家裏拿出一些破舊的鍋子、廢鐵、雞毛、頭髮等等的東西賣給「收買佬」，是真的！現在想起來，確是一個奇景！而其中一樣被收買而又肯定家家都必定有的，就是平時吃柑剝下來的那層果皮。通常大人都會將吃剩下的柑皮用繩子串成一串串，然後掛起來吹曬乾透之後，就把它儲存起來，等到有「收買佬」來的時候就可以賣出去。那時候，我根本不知道那些果皮到底有甚麼用，只知道，若我乖乖的把柑皮儲起來，便可以賺到一元幾毫買東西吃。直至長大後才知道，那些隨著我年紀一起增長，由果皮變成的陳皮，原來是可以那麼值錢及有食療價值的。

無錯！陳皮就是我們平時吃柑橘的外皮，經過陽光照曬風乾之後，便變成了陳皮，而存放時間愈久的陳皮，食療效用就愈佳，而其中又以廣東新會所出產的陳皮最為著名，以大片、香氣濃郁為上品，亦被稱為廣東三寶：陳皮、老薑、禾稈草之一。

據中醫理論，陳皮具有理氣化痰、健脾的作用，對消化不良、食慾不振、嘔吐、咳嗽痰多等有明顯的食療功效。因此，陳皮亦非常普遍地被應用在烹調上，例如煲老火湯的話，我必定下片陳皮，廣東人的冬瓜老鴨湯、清補涼、紅豆沙、薑茶糖水等，如果沒有了陳皮，這些湯水便會黯然失色了。還有，當你煲粥、蒸魚、蒸煮排骨或其他的肉類、肉餅，只要下些陳皮，整道餸菜便如同畫龍點睛，馬上生色不少了。我其中一道的創作菜代表作「香酥九製陳皮骨」，陳皮便是這道菜的靈魂。除此之外，陳皮還能辟腥去臭，簡直是廚房不可缺少的法寶之一。

其實，自己在家曬製陳皮亦相當簡單，只要將柑皮串起來掛在窗口或用筲箕盛著放在露台或窗台，在陽光下曬至硬身及完全乾透（約四至五星期），然後再放入焗爐，以攝氏60度至80度烘約四至五小時，以確保去除所有水分，涼卻後便可以放入存氣的瓶內保存，一年後便可以使用了。在冬天，正是柑的當造季節，天氣風高又乾燥，正是曬製陳皮最好的時候了。

陳皮を自分で作る 273

Making your own dried tangerine peel 273

- 要揀選大個皮厚的柑皮，曬製出來的陳皮才夠香及耐存，細隻皮太薄的柑皮除了不夠香之外，曬乾後亦較易碎裂。

- 陳皮若是處理不當亦很易發霉，若買回來的陳皮如果發潮變軟，可以先用布擦去霉漬，再翻曬及用焗爐焙乾，然後放入存氣的瓶子再保存，陳皮最好存放在通風、乾燥及陰暗的地方。

- 陳皮使用前先用清水浸軟，洗淨，然後用小刀將皮內的瓢刮去，因果瓢較濕熱。陳皮每次用一至兩小片便夠，太多會令同時烹調的食物帶有苦味。

傳統廣東甜酸泡菜

伝統の広東ピクルス ...274

Traditional Cantonese pickles ...274

材料

白蘿蔔　1條
小黃瓜　2條
甘筍　2條
中型紅辣椒　1-2隻
蒜頭　3-4粒

醃料

鹽　3茶匙
白醋　1公升
砂糖　800克

做法

1. 材料全部洗淨。白蘿蔔刨皮切粒，小黃瓜去籽切條，甘筍切片，辣椒切圈，蒜頭切片，然後放入容器內，下鹽撈勻放置一天讓它出水。

2. 白醋加入砂糖，以小火煮至砂糖溶解後關火，待涼卻後備用。

3. 第二天將排出的水分倒掉，並用手揸乾材料的水分。

4. 最後將已揸乾水分的材料放入容器內，加入已溶解砂糖的白醋撈勻，泡醃約兩天便可以食用了，吃時可灑些炒香芝麻伴食，更加滋味。

貼士

這是最為簡單易做的傳統廣東式泡菜，你可以再混合一些自己喜歡的蔬菜一同泡醃，如椰菜、沙葛、豆角等，同樣美味。

冬 _ Winter

梅醋醃蓮藕

レンコンの梅酢漬け...275

Pickled lotus root in green plum vinegar...275

🍀 材料

蓮藕　約1公斤
梅醋　500毫升
砂糖　適量

🍀 做法

1. 蓮藕洗淨，刨皮，切片，放入滾水中灼約數十秒，撈起馬上浸冰水至完全涼卻，盛起瀝乾水分備用。

2. 梅醋加入適量砂糖，調至自己喜愛的濃度，將蓮藕放入容器內，加入梅醋撈勻，泡醃約一天便可食用。

貼士 🍀

整條蓮藕的形狀是一節一節，由粗至幼地生長，較粗的部分是最成熟的，口感較綿，味道較濃，適合用來煲湯；而幼身部分則較嫩，口感爽脆，適合用來炒及做泡菜。蓮藕基本上可以生吃，但因為它生長在水裏及污泥之中，易有細菌，所以用來做泡菜的話，必須要用滾水燙過以確保安全(用滾水燙過的蓮藕撈起後馬上以冰水泡浸，便不會影響其爽脆口感)。

醬油小黃瓜

キュウリの醬油漬け .275
Soy-pickled baby cucumbers .275

❧ 材料

温室小黃瓜　1公斤
鹽　2茶匙

❧ 醃料

美極鮮醬油　150克
米酒　100克
冷開水　100克
砂糖　100克
甘草　2片

❧ 做法

1. 小黃瓜洗淨，抹乾水分，切成小段，下鹽撈勻，放置一晚讓它出水。
2. 第二天將小黃瓜排出的水分倒掉，再用手揸乾水分；攪勻醃料，放入小黃瓜，泡醃約兩天便可以食用了。

貼士 ❧

1. 醃料中的水一定要使用煲滾後涼卻的冷開水，若用生水的話，醃菜很易變壞及發霉，因生水含有很多雜質及微菌。
2. 醃製好的小黃瓜可以當成涼菜直接食用，亦可成為粥、粉、麵、飯的配菜，用來拌炒各類菜式亦非常爽口美味。

酸辣千層大白菜

酸辣(サンラー)の千層白菜...276

Sour and spicy Tianjin white cabbage pickles...276

❈ 材料
大白菜　1棵(約2公斤)
中型青、紅辣椒　各3-4隻
(視乎自己愛辣的程度加或減)
蒜頭　3粒
鹽　3茶匙

❈ 醃料
白醋　1公升
砂糖　800克

❈ 做法
1. 白醋加入砂糖，以小火煮至砂糖溶解後熄火，待涼卻後備用。
2. 大白菜洗淨，瀝乾水分，對切一開為二；青紅辣椒、蒜頭洗淨，抹乾，切片；全部放在容器內，灑上鹽抹勻大白菜內外，然後用手將大白菜輕輕搓揉，至大白菜變成軟身，然後放置一天讓它出水。
3. 第二天將排出的水分倒掉，並揸乾水分，全部材料放在大盤內，注入已溶解砂糖的白醋，撈勻，泡醃約兩天便可以食用。

貼士 ❈
1. 用整棵大白菜醃製可以達到賣相精美的效果，若覺得難以處理，可以將大白菜切片再醃製，這樣亦不會影響味道及口感。
2. 其他貼士可參考第121頁傳統糖醋醃青木瓜。

冬 _ Winter

海南雞辣椒醬醃椰菜花

カリフラワーの海南鶏チリソース漬け...277

Pickled cauliflower with chilli sauce
for Hainan chicken...277

美味的海南雞飯很多人都吃過，除了軟滑的雞肉和那碗用浸雞水、雞雜、雞油煮成的白飯之外，調配得好的黑豉油、薑葱茸和那碟酸酸辣辣的辣椒醬更是海南雞飯的靈魂。而那碟辣椒醬裏面除了辣椒之外，還有蒜茸、白醋、砂糖等美味材料，很適合用來醃製泡菜，於是我便嘗試用它來做泡菜，效果相當不錯。

🍀 材料

椰菜花　1個(約700克)
鹽　2茶匙
炒香芝麻　少許
麻油　適量

🍀 醃料

海南雞辣椒醬　1瓶(230克)
砂糖　2湯匙

🍀 做法

1. 將椰菜花洗淨切成小朵，下鹽撈勻，放置一天讓它出水，期間要翻動及輕輕搓揉數次，這樣會令水分較少及較硬身的椰菜花鬆軟些。

2. 第二天將椰菜花飛水後馬上再浸冰水，至椰菜花冷卻後用手捏乾水分放入盆內，加入醃料撈勻醃約一天便可以食用了。

3. 吃時可灑上適量炒香芝麻及少許麻油伴吃。

貼士 🍀

1. 由於每款牌子的海南雞辣椒醬的酸辣度不一，可因應個人口味再調整砂糖的份量，甚至可以再加些白醋或青檸檬汁。

2. 海南雞辣椒醬可於各區賣椰汁、咖喱、香料的東南亞雜貨舖買到，除了醃椰菜花之外，當然亦可用來泡醃其他你喜歡的瓜菜。

糟滷醉鮮鮑

糟滷（酒粕）酔っぱらいあわび ...278

Wine-Marinated abalones in distiller's grain sauce ...278

我很喜歡吃鮑魚，喜歡它豐滿的口感及鮮美得令人忘形的味道，我嘗試過用很多不同的方法去炮製鮑魚，以滿足自己對鮑魚的貪婪慾望。我發現除了以燜煮的方式處理鮑魚外，用醃漬的方法做出來也極之好味道。這次我用鹹鮮的糟滷，略帶醉意的花雕酒及麻麻香香的花椒油來泡醃澳洲鮑魚，除了鮮味之外，還多了一份葱味的新鮮感！

🎋 材料

澳洲新鮮鮑魚　2隻(每隻約600克)
薑片　約5片
葱　3-4棵
雞湯　2公升
冰糖　200克

🎋 醃料

上海糟滷　1枝(約500毫升)
紹興花雕酒　350毫升
煮完鮑魚的雞湯　150毫升
冷開水　100毫升
花椒油　50毫升

🎋 做法

1. 鮑魚去殼，去內臟，擦洗乾淨後以薑葱水飛水。

2. 鮑魚以雞湯、冰糖慢火煲約4小時至腍身，涼卻後備用。

3. 醃料撈勻，放入鮑魚泡醃一晚使其入味。第二天將鮑魚撈起切片，便可以享用鮮甜味美的糟滷鮑魚了。

貼士 🎋

1. 鮑魚至少要醃8小時才夠入味，但亦切勿泡醃超過一天，否則便會太鹹。若不是即時食用，可充分泡醃後將鮑魚撈起，再獨立存放。

2. 燜煮鮑魚時不要加入有鹹味的調味料，否則鮑魚體積會縮小及不易燜腍。

沙爹 XO 醬

在我記憶之中，XO醬開始在香港流行應該是二十多年前了，這是一道相當本土口味的香港式醬料，至於到底誰是真正的原創者，至今都有幾個不同的版本及說法，今次暫且不去探究。但無可否認，XO醬是一款相當受香港人歡迎的醬料，它不單可以配襯各式小菜、小食、粥、粉、麵、飯，亦可烹調炒煮各樣菜式，而且味道惹味討好，可謂是用途廣泛，廚房必備的醬料之一。基本上，現在各大酒家及餐廳，都有自己配方的XO醬，甚至已經包裝成自家產品，甚至暢銷海外了。

其實，在家自己炒煮XO醬亦不是大家想像中那樣複雜。一般XO醬的主要材料大致為蝦米、瑤柱、火腿、辣椒、乾葱、蝦子等，但經過多年的演變及改良，今時今日的XO醬無論材料及口味都已經很多元化，有些更加入鹹魚、銀魚仔、三文魚、海味、菇菌類等，甚至有專為素食者而設的素XO醬。而我本人亦很喜歡XO醬，我亦嘗試過以多種不同材料及不同份量比例做過多款口味的XO醬，除了可以配以多類小菜，有南洋風味的XO醬，希望大家喜歡這款帶冬天吃火鍋用來做配醬效果相當不錯！

サターXO醬 279
Satay XO sauce...279

❧ 材料
瑤柱　100克
蝦米　100克
指天椒　50克
乾葱頭　300克
蒜頭　150克
油　700毫升

❧ 調味料
砂糖　20克
沙爹醬　120克

❧ 做法
1. 瑤柱、蝦米洗淨，浸水一晚至軟身；瑤柱撕成絲，蝦米用石舂舂成蝦米鬆，指天椒洗淨切細圈，乾葱頭洗淨去皮切片，蒜頭剁成蒜茸，備用。
2. 鑊中燒熱油，以小火爆香指天椒、乾葱片及蒜茸，至材料收水開始變金黃色時下瑤柱及蝦米鬆炒拌，至全部材料開始乾身及呈金黃色，此時醬料亦開始起泡，這時便代表醬料已差不多完成了，最後下沙爹醬及砂糖炒拌均勻及油再滾起時即可關火，盛起涼卻後便可放入經消毒處理的容器保存，即成私房秘製沙爹XO醬了。

貼士 ❧
1. 一般製作XO醬的蝦米都是用整隻的，除了喜歡它的口感之外，在賣相上見到原隻蝦米，感覺上真材實料一些。但我今次將蝦米用石舂舂成蝦米鬆來做XO醬，目的是讓蝦米的味道更能與其他材料完全溶合，令「醬」的質感更實在一些，吃起來亦另有一番滋味。
2. 炒XO醬一定要小火慢炒，這樣材料的水分才夠時間慢慢釋出，完成後材料仍可保持鬆軟口感之外，因為水分少，保存時間亦會較長一些。

冬 _ Winter

大根の麻辣(まーらー)漬け ...280
Radish pickles with Sichuan peppercorns ...280

材料

白蘿蔔　約2.4公斤
鹽　2茶匙

麻辣醃料

蒜頭　3粒(切片)
花椒　5湯匙
豆瓣醬　8湯匙
辣椒油　3湯匙(視乎個人吃辣程度)
鹽　2茶匙
砂糖　450克
白醋　700毫升
麻油　4湯匙

做法

1. 蘿蔔洗淨，抹乾水分，刨皮並切成約兩吋長左右條狀，放入盆內，加入2茶匙鹽與蘿蔔條撈勻，醃約四至五小時。

2. 四至五小時後，蘿蔔條會排出水分，將排出的水分倒掉，並用手搾乾蘿蔔。

3. 將所有麻辣醃料拌勻，放入蘿蔔撈勻，泡醃一至兩天後便可以食用。

貼士

1. 醃製泡菜要吃起來爽脆，一定要先用鹽醃過，讓它出水，搾乾，然後才放入醃料泡醃，因為蔬菜、瓜、根莖本身含有水分，若不搾乾的話，水分會在泡醃過程中排出，因而稀釋了醃料的濃度，影響了泡醃出來的味道及爽脆口感。

2. 泡醃期間要將蘿蔔及醃料再撈勻兩至三次，使蘿蔔能更均勻入味。

味噌醃鹹蛋黃

アヒル卵黄の味噌漬け ...281
Miso-salted egg yolk ...281

大家熟悉的日本味噌，都被認為是主要用在烹調各類麵豉湯類，其實用在其他菜式的調味或醃製食物亦非常普遍。這次我用味噌來醃製鹹蛋黃，你會發現，除了口味很有新鮮感之外，還會帶來不一樣醃製風味！

🍀 材料

鴨蛋　約9個

🍀 醃料

日本味噌(低鹽)　1盒(約600克)
米酒　2湯匙
砂糖　5湯匙

🍀 做法

1. 醃製之前，先將鴨蛋放在雪櫃約半天，讓蛋黃的狀態較為凝固，易於處理。

2. 醃料攪勻，取一有深度的容器，先鋪上一層醃料，再鋪上一層紗布，取鴨蛋較大的底部在紗布上壓出一個可以容納蛋黃的凹洞。打開鴨蛋，小心取出蛋黃，放在凹洞內，將紗布的另一邊覆蓋上蛋黃，在上面亦小心均勻地抹上一層醃料，再用保鮮紙封好，放入雪櫃。盡量不要移動它，放置三個星期左右，蛋黃會變硬凝固成固體狀，這時便可以使用了。

貼士 🍀

1. 醃製好的味噌鹹蛋黃，你可以將它蒸熟，用來配粥、鹹豆漿之類；亦可以代替傳統鹹蛋黃烹調各類菜式，如蛋黃蝦、蛋黃蟹、蛋黃涼瓜、蛋黃冬筍等等……

2. 在最後抹上醃料在紗布上的時候，不能太厚，(約5毫米便夠了)，太厚的話重量便很容易將蛋黃壓破。

3. 蛋黃在放入去凹洞之前要盡量將蛋白拿走，否則太濕的話會影響蛋黃醃製出來的效果。

4. 市面上有很多不同風味的日本味噌，你可以因應自己的口味而選擇，不同的味噌做出來亦有不同的口味，但最好揀選一些低鹽的味噌，因為是直接醃製蛋黃，傳統的味噌醃製出來可能會過鹹。

冬 _ Winter

味噌卵黄のソフト·シェルクラブ ..282
Miso-salted egg yolk crusted soft-shell crabs ..282

味噌蛋黃軟殼蟹

🍀 材料

軟殼蟹 4隻
味噌鹹蛋黃 8個
生粉 適量（炸蟹時上粉用）

🍀 脆漿粉

麵粉 150克
泡打粉 1克
生粉 10克
清水 適量

🍀 做法

1. 軟殼蟹解凍後去腮，洗淨並揸乾水分，剪開，一開為二；味噌鹹蛋黃蒸約10分鐘至熟，再用义壓碎備用。

2. 脆漿粉撈勻，慢慢加入清水，至濃稠度可以用手指掛起及提起時能成直線滑下，再加入蛋黃碎攪勻成脆漿粉。

3. 軟殼蟹先撲上適量生粉，再均勻沾上調好的脆粉漿，燒熱油約180℃，炸至金黃香脆即可享用。

貼士 🍀

1. 脆粉漿不能調得太稀，太稀的話蛋黃便黏不住粉漿，容易在炸的過程中飛散，另外蛋黃亦不宜壓得太碎，否則口感不明顯。

2. 處理酥炸的菜色若要做出香脆酥化的效果其實有些小要訣，通常餐廳廚師會在炸食物時會先用中火炸，當食物炸至開始金黃色時，便將火加大，令油溫昇高再炸10數秒左右，目的是讓高油溫將食物的水分逼出，這樣炸出來的食物便會鬆化香脆，食物咬下去時亦不會感覺太油膩。但在一般家庭廚房我們可以先將快炸好的食物撈起，然後擱置幾分鐘後，再將油燒熱炸多一次，這樣也能做到相同的效果。

3. 一些大型日式超市所出售的現成脆漿粉其效果也不錯，若貪方便的話，可以代替自己調配。

冬 _ Winter

味噌醃鱈魚

タラの味噌漬け ...283
Grilled cod in miso sauce ...283

🌼 **材料**
鱈魚扒　兩件

🌼 **醃料 (A)**
日本清酒　約2湯匙

🌼 **醃料 (B)**
味噌　900克
砂糖　250克
味醂　150克

🌼 **做法**
1. 鱈魚去鱗洗淨，用廚紙吸乾水分，再用清酒均勻搽滿魚扒底面，備用。
2. 將醃料(B)拌勻，取一有深度的碟，先在碟底均勻抹上一層醃料，鋪上紗布，然後擺上鱈魚，再將紗布另一邊覆蓋上鱈魚，在上面亦均勻抹上醃料，醃約一晚。
3. 第二天將鱈魚放於烤盤，放入焗爐以250℃焗約6-7分鐘至鱈魚金黃微焦帶脆即成，吃時可配以沙律醬及灑上青檸檬汁伴吃。

貼士 🌸

1. 在碟上塗抹醃料時厚度一定要平均，不要凹凸不平，這樣鱈魚才能均勻入味。

2. 鱈魚不要醃超過一天，否則會太鹹，若不是即時烹調，可先將鱈魚從醃料取出，再獨立存放，以免醃漬時間太長而過鹹。

3. 鱈魚扒不要切得太薄，否則完成後肉質很易鬆散。此道菜除了鱈魚之外，亦可以其他魚類代替，各有風味。

[Ingredients]

900 g fresh kumquats
900 g coarse salt
1/4 cup distilled water or white vinegar

[Method]

1. Rinse the kumquats and drain well. Leave them under the sun for 2 to 3 days until their skin turns wrinkly and dry. Remove the stems and wipe down the kumquats with a dry cloth.

2. Lay a layer of coarse salt in a sealable wide-mouth glass container. Top with a layer of kumquats. Repeat until all kumquats and salt are used. Make sure the top-most layer is coarse salt. Drizzle with distilled water or white vinegar.

3. Seal the container. Write down the production date. Leave them in a cool dry place away from the sun (such as inside a kitchen cabinet). Water will be drained out of the kumquats after 1 week and the salt starts to dissolve. The kumquats turn from bright orange into light brown and then dark brown.

[Tips]

1. There are several reasons why the kumquats are dried under the sun for a few days before being salted. First, the ultraviolet light from the sun helps kill bacteria. Secondly, the salted kumquats tend to last longer when the moisture in the skin is removed in prior. Third, as the water content of the kumquats is reduced, they produce a stronger citrus flavour and aroma after being salted.

2. Whenever you pickle or salt any food, always use a wide-mouth clear glass container. This way, you can put the food into or take the food out of the container easily. You also get to see how the food changes in colour and texture. The food also lasts longer in a glass container. Because of the high concentration of salt and acid in the pickling process, plastic containers could cause problems. Finally, always sterilize the container in boiling water and then dry it completely before using it for salting or pickling.

3. Before you salt the kumquats, make sure they are dry on the outside. Any uncooked water remaining on the skin would make them susceptible to moulding.

4. Salted kumquats are high in salt content and they last for a long time at room temperature, just like aged wine. In fact, the longer they stand, the richer and tastier they become. There are people out there keeping their salted kumquats for decades and treat them like expensive wine.

5. Most kumquat plants we get for decoration around Chinese New Year are treated with spray-on flower food or preservatives to brighten their colours and extend their life. Thus, kumquat fruits from these plants may carry some harmful chemicals. Thus, before you salt them, make sure you rinse them thoroughly.

[材料]

新鮮な金柑...約900g
粗塩...約900g
1/4杯の 蒸留水又は白酢

[作り方]

1. 新鮮な金柑を洗い、水気を取り、天日に2～3日干し、金柑の皮が縮んで初め、皺が少し出る程度、ヘタを取り、ふきんで金柑をきれいに拭き取る。
2. 広口で密閉できるガラス容器に粗塩、金柑を交互に入れる。最後1/4杯の蒸留水又は白酢を加える。
3. 容器にいっぱいまで入れてから密閉して、日付を記入し、冷暗乾燥と直射日光が当たらない場所に置く(台所に置くのがお勧め)。約一週間後、金柑からの水分が除々に出るによって、粗塩が溶かす、さらに金柑の鮮やか黄色からどんどん色がくすんできて、茶色っぽくなってきました。

[ポイント]

1. 金柑を漬ける前に天日に2～3日干しの原因、一は太陽の紫外線で殺菌する。二は令柑の表皮から水分を取られ、こうしたら、漬けた金柑が腐りにくく、長時間に保存できる。三は金柑の表皮から水を取られたら、金柑の味が濃くなり、漬けた金柑の香りがもっと強く出せる。
2. 全ての漬け物は、広口と透明な容器を使用し、漬け物を取り出しやすく、漬けたもの、その変化の様子をいつでも見られる。そして、長時間に保存できる。高濃度の塩分と酸性のある漬物はプラスチック容器には適用しない。容器に漬け物を入れる前に、熱いお湯を流して乾いてから使う、これは消毒が効果的となる。
3. 金柑を漬ける前に、必ず金柑が乾くなるまで確認し、生水が付けると、かびの発生しやすくなる。
4. 漬け上がった金柑は塩分濃度が高いので、長時間保存でき、古い酒のように長く保存するほど、コクが出る。名酒のように何十年でも長く保存する人がいるんだ！
5. 中国の正月を迎える際に、キンガンが多く使われ、市販が金柑を光って見え、長くキープできるように、金柑の皮の表面には化学剤のようなものを塗布されたので、必ず金柑をしっかり洗ってから漬けること。

[Ingredients]

1 grey mullet (about 900 g)
6 to 8 salted kumquats
1 tbsp fermented soybean paste
3 to 4 slices ginger
finely shredded spring onion
2 tbsps oil

[Method]

1. Dress the fish and rinse well. Put sliced ginger on a plate. Put the fish on top. Slice the salted kumquats and arrange on top of the fish. Spread fermented soybean paste evenly over the fish.
2. Boil water in a wok or steamer. Steam over high heat for 8 minutes. Remove from heat. Put shredded spring onion on the fish. Heat oil until smoking hot. Pour over the spring onion. Serve.

[材料]

ボラ...一匹（約900g）
金柑塩漬け...6〜8粒
中国味噌...1大さじ
しょうが（片きり）...3〜4片
ねぎ（せん切り）...適量
油...2大さじ

[作り方]

1. ボラをおろして洗い、しょうがの薄切りをボラの上にのせ、金柑の塩漬けを薄切り、ボラの上にのせ、そして中国味噌をのせる。
2. 水を沸騰してから、大火で8分ほどに蒸し、取り出し、ボラの上にねぎをのせ、香ばしく炒めた油をボラにのせて召し上がる。

[Ingredients]

150 g salted kumquats (about 16 kumquats)
1 kg fresh kumquats
200 g sugar
2 tbsps lime juice
1/2 cup honey
800 ml water

[Method]

1. Seed the salted kumquats and finely chop them. Set aside. Rinse and drain the fresh kumquats. Chop them up and seed them. Add sugar and lime juice. Mix well. Leave it overnight to drain the moisture out of the fresh kumquats.
2. Add water and salted kumquats to the fresh kumquat mixture. Bring to the boil over high heat. Turn to low heat. Cook while stirring constantly for about 20 minutes until the liquid starts to bubble, looks satiny and dense. Turn off the heat. Leave it to cool completely. Transfer into bottle and seal well.

[Tips]

1. Make sure you seed the kumquats. Otherwise, the marmalade will taste bitter and you'd have to spit the seeds out when eating it.

[材料]

金柑の塩漬け...150g（約16粒）
新鮮な金柑...600g
砂糖...200g
ライムジュース...2大さじ
はちみつ...半杯
水...800ml

[作り方]

1. 金柑塩漬をみじん切り、種を取り出し、新鮮な金柑を洗い、水気を取り、みじんきり、種を取り出し、砂糖とライムジュースを混ぜ合わせ、水を出させるため一晩置いておく。
2. 翌日に水を加え、大火で沸騰しから弱火に材料を撹拌しながら約20分に煮、泡を立てジャーム状になるまで火を止め、ジャムを盛り上げ、冷めてから、ボトルに保存し、ゆっくり召し上がる。

[ポイント]

1. 必ずに金柑の種を取り出すこと。そうしないと、ジャームの味が苦くなり、召し上がるときに種を噛んでしまうと、せっかくの食感を壊してしまう。

Homemade kumquat wine
自家製金柑酒..........P24

[Ingredients]
3.375 kg fresh kumquats
2.7 kg Guangdong rice wine
1.8 kg rock sugar

[Method]
1. Rinse the kumquats and remove the stems. Wipe dry. Put kumquats, rice wine and sugar into an airtight glass bottle. Seal well. Leave them in a dry cool place away from the sun for 1 year. Serve.

[材料]
新鮮な金柑...3.375kg
広東米酒...2.7kg
氷砂糖...1.8kg

[作り方]
1. 金柑を洗い、ヘタを取る、水気を取り、密閉ガラス容器に米酒、氷砂糖と一緒に入れ、冷暗乾燥と直射日光が当たらない場所に約一年間を保存して、召し上がる。

Osmanthus kumquat wine
桂花金柑酒..........P25

[Ingredients]
1.575 kg fresh kumquats
3 bottles aged osmanthus wine (750 ml per bottle)

[Method]
1. Rinse the kumquats and remove the stems. Wipe dry. Put kumquats and aged osmanthus wine into an airtight glass bottle. Seal well. Leave them in a dry cool place away from the sun for 6 months. Serve.

[材料]
新鮮な金柑...1.575Kgs
桂花陳酒...3ボトル(ボトル 750ml)

[作り方]
1. 金柑を洗い、ヘタを取る、水気を取り、密閉ガラス容器に桂花陳酒を入れ、冷暗乾燥と直射日光が当たらない場所に約半年間に置き、召し上がる。

217

**Homemade green plum wine (It's the plum season again.
The ripe plums are so tempting...**

自家漬け梅酒（梅の季節到来！うわぁ～かなり誘惑だ...） P27

[Ingredients]

2.7 kg green plums
2.25 kg rock sugar
2.7 kg rice wine

[Method]

1. Rinse the plums and drain well. Leave them to air dry completely. Then use a toothpick to remove the stems. Prick one or two holes on each plum so that the wine can be infused with plum juice more easily.
2. Put the plums, rock sugar and rice wine into a sealable wide-mouth transparent glass container.
3. Seal well. Write down the production date. Leave them in a cool dry place away from the sun. Swirl the bottle every once in a while to speed up infusion until all rock sugar dissolves. Then you can just leave it still for 3 months. By then, the wine should be golden and clear with a fruity smell and sour-sweet palate.

[Tips]

1. When you make any fruit wine, the alcohol content of the rice wine should be at least 30%. It's because the fruit flavour is extracted by the alcohol. Wine with too little alcohol content is not efficient enough. You may use Yu Bing Shao, Guangxi rice wine, double-distilled or triple-distilled rice wine, Jiujiang rice wine or even Japanese sake.
2. For making wine, pick green plums that are ripe and yellow in colour. The plum wine will end up darker in colour with a stronger plum taste. If you pick green and unripe plums, the plum wine will end up lighter and clearer in colour, with a sourer taste.

[材料]

青梅...2.7kg
氷砂糖...2.25kg
米酒...2.7kg

[作り方]

1. 梅を洗い、水気を取り、梅は完全に乾くなってから、なり口のホシを竹串などで丁寧に取り除く。梅汁をでやすいため、1～2小穴を刺す。
2. 広口で密閉できるガラス容器に梅、氷砂糖、米酒を交互に入れる。
3. 蓋を閉め、日付をつけ、冷暗乾燥と直射日光が当たらない場所に保存し、時々容器を揺り動かして、氷砂糖を完全に溶けてから静止した状態で3ヶ月を経ってから、あっさりとした味、澄んだ金色、上品な梅の香り、爽やかな酸味の自家漬け梅酒ができました。

[ポイント]

1. 一般的な果実酒は、30度以上の米酒を使用し、アルコールが高ければ高いほど果実の香りがもっと美味くなる。例えば、中国米酒（玉冰焼、広西米酒、雙蒸、三蒸、九江米酒など）と日本清酒とも使用できる。
2. 黄色っぽいと熟した梅子を選んだら、梅酒の色は深くなり、味が濃くなる。逆に青々した硬い梅子を選んだら、漬けた梅酒の色は浅くなり、梅の酸味が強くなる。

218

Muscovado plum wine
濃醇黒糖梅酒..........P30

[Ingredients]

2.7 kg half-ripe or fully-ripened green plums
2.7 kg rice wine
2.25 kg muscovado

[Method]

1. Rinse the plums. Drain. Leave them to air dry completely. Use a toothpick to remove the stems. Then prick 1 or 2 holes over the skin for better infusion of the plum juice.
2. Place the plums, rice wine and muscovado into a sealable wide-mouth transparent glass container. Seal the container well. Leave it in a cool dry place away from the sun. Shake the container once in a while to speed up infusion. After the muscovado dissolves, leave it still for 3 months. Serve.

[Tips]

1. This wine has a great fruity sweetness. Both the wine and the plums that have been steeped in it carry a lovely stickiness and caramel-like flavour which is characteristic of muscovado. Thus, both the wine and the candied plums go well with various desserts. I prefer using half-ripe or fully ripened plums for their richer plum flavour and fragrance which yield a richer wine at last.
2. Besides regular muscovado, you may also use Japanese Kurozatou from Okinawa. The wine will end up even richer and tastier, yet also cloudier in appearance. On the other hand, using muscovado will yield a clearer wine, but it won't taste as strong and rich as Kurozatou.

[材料]

半熟または完熟梅子...2.7kg
米酒...2.7kg
黒砂糖...2.25kg

[作り方]

1. 梅を洗い、水気を取り、完全に乾かせる。竹串でヘタをとる。実にぶすぶすと穴を開け、エキスを出やすくする。
2. 広口の密閉できる透明なガラス容器に梅、米酒、氷砂糖を入れ、蓋を閉め、日付をつけ、冷暗乾燥と直射日光が当たらない場所に保存し、時々容器を揺り動かして、黒砂糖を完全に溶けまで、静止した状態で３ヶ月放置してから召し上がる。

[ポイント]

1. 黒糖梅酒は甘くて香りが良い、黒糖梅酒もしくは梅酒に浸された梅はデザートによく合わせ。半熟又は黄色になった梅、熟した梅と黒砂糖に混ぜ合わせて、梅酒の香しい香りともっと濃淳な味になる。
2. 伝統的な黒砂糖で漬ける。例えば、日本の著名な沖縄産黒砂糖は味がもっと香ばしく、もっと濃くなる。しかし、漬けた酒の色が濁りになる。もし普通の黒砂糖で漬けた梅酒はちょっとさっぱりする。しかし、その濃厚的な味と香りは伝統的な黒砂糖に及ばない。

Green plum wine jelly
梅酒ゼリー..........P32

[Ingredients]

1 litre homemade green plum wine
50 g gelatine leaves (about 10)

[Method]

1. Soak the gelatine leaves in iced water for 10 minutes until soft. Squeeze dry.
2. Heat the green plum wine to 60 – 70℃. Put in the gelatine leaves and stir until they dissolve. Strain the resulting mixture to remove any solid bits and impurities.
3. Pour the plum wine and gelatine mixture into dessert cups or a flat tray. Refrigerate until set. Serve.

[材料]

自家製梅酒...1リットル
板ゼラチン...50g（約10枚）

[作り方]

1.板ゼラチンを氷水で約10分ふやかす、水気を切って使用。

2.梅酒を60〜70度ぐらいまで加熱し、板ゼラチンを入れ、完全に溶けるまで撹拌し、漉し器でゼラチンの溶け残りを漉す。

3.梅酒をデザート容器或る皿に入れ、冷めてから冷蔵庫に入れ、ゼリー状になってから出来上がる。

Green plum honey
蜂蜜梅干..........P38

[Ingredients]

1 kg half-ripened or unripe green plums
1.5 kg honey

[Method]

1. Rinse the plums. Drain well. Remove the pits of the plums with a bamboo skewer. Prick one or two holes on each plum so that the honey can be infused with the plum juice more easily.
2. Transfer the plums into a container. Pour in the honey and seal the container well. Leave it in a cool dry place in the shade away from the sun for about 3 months. The plum juice should be drawn out of the plums then to blend well with the honey. The plums would shrink and turn wrinkly. The honey is ready to be served then.

[Tips]

1. To serve the honey, simply dilute it with water. I prefer to serve it iced because the natural nutrients in honey are likely to be destroyed by heat. You can also use it as a syrup or topping for ice cream, dessert or cakes.

[材料]

半熟又は青梅...1000g
蜂蜜...1500g

[作り方]

1.梅を洗い、水気を取り、竹串でヘタを取り、梅に1〜2箇所穴を開けて、梅のエキスを出やすくする。
2.梅を容器に入れ、蜂蜜を加え、蓋をする。乾燥とした直射日光が当たらない冷暗所に保存し、約三ヶ月に梅のエキスと蜂蜜が混ざり合うとともに、梅の実が萎んできたら出来上がる。

[ポイント]

1.蜂蜜梅干を水で割って飲む。冷たく冷やしたら風味がもっと良い。しかし、お湯で割ると蜂蜜の持つ栄養素を破壊してしまう。蜂蜜梅干は各種のアイスクリーム、デザート、ケーキに加えても美味しい。

Homemade green plum vinegar
自家製梅酢 P36

Drink vinegar for health

Green plum vinegar, as its name suggests, is made with green plums soaked in vinegar. It has the therapeutic effects of both plums and vinegar. Not only does it taste delicious, it also quenches thirst, eases physical exhaustion, expels Heat, stops cough and benefits the Stomach. It also neutralizes the acidity or alkalinity of food.

[Ingredients]

1.5 kg half-ripened or ripe green plums
1.5 litre white vinegar
800 g rock sugar

[Method]

1. Rinse the plums. Leave it to air dry completely. Use a toothpick to remove the stems. Then prick one or two holes on each plum so that the vinegar is easily infused with the plum juice.
2. Place the plums into a container. Pour in the white vinegar and rock sugar. Seal the container well. Leave it in a cool place out of the sun. Swirl the container every once in a while to speed up infusion until all rock sugar dissolves. Leave them to soak for at least 6 months.

お酢でヘルシー

梅で漬けた梅酢は、両方の効用を持ち、いわゆる一挙両得となる。梅酢は美味しいだけではなく、のどの渇きをいやす、疲労回復、解熱、のどが渇くない、疲労を止め、解熱、鎮咳、胃の調子を整える。更に酸性とアルカリ食べ物を中和できる。

[材料]

半熟または完熟梅...900g
白い酢...900g
氷砂糖...800g

[作り方]

1. 梅を洗い、完全に水気を取り、ようじでヘタを取り、梅に一つ小穴を指し、梅汁を出やすくさせる。。
2. 梅は容器に入れ、白い酢と氷砂糖を入れ、蓋を閉め、日付を記入、冷暗乾燥と直射日光が当たらない場所に保存し、時々容器を揺り動かして、約6ヵ月後、梅酢を召し上がる。

Homemade muscovado green plum syrup
自家製黒糖梅シロップ..........P40

[Ingredients]

2 kg half-ripened green plums
2 kg muscovado (or Japanese Kurozatou)
1/2 cup rice vinegar
3 tsps coarse salt

[Method]

1. Rinse the plums well. Drain. Remove the stems with a toothpick. Then prick 1 or 2 holes on the skin so that the syrup can be easily infused with plum juice.

2. In a large tray, pour in a layer of muscovado. Top with a layer of plums.

3. Repeat step 2 until the tray is full. Make sure the top-most layer must be muscovado.

4. Sprinkle coarse salt on top. Gently press the mixture to pack the plums and muscovado firmly. Without space between them, the infusion can take place more quickly. The plum juice will also dissolve the muscovado more quickly.

5. Pour rice vinegar on top. Seal the tray with cling film. Write down the production date.

6. Leave them for 1 week. Liquid will ooze out ot the plums and turn the syrup yellow. The muscovado will dissolve slowly. At this point, stir the mixture to mix well.

7. By the second week, more muscovado dissolves into the plum juice. Now, the plums are soaked in syrup. The plums turn softer and lighter golden in colour.

8. At this point, you may seal the plums and syrup. Pour the mixture into an airtight, wide-mouth, transparent glass container. Seal well and write down the production date. Keep it in the fridge and allow the infusion to go on.

9. After 2 months, all sugar must have dissolved and the muscovado plum syrup is done. Similar to a plum wine, it gives a fruity floral plum fragrance. But this syrup is not as complicated as plum wine, which takes a long time to mature.

[材料]

半熟梅子...1200g
黒砂糖...1200g
米酢...半杯
粗塩...3小さじ

[作り方]

1. 梅を洗い、水気を取り、くぼみに竹串などで刺すようにして梅のヘタを取り、実にぷすぷすと穴を開け、エキスを出やすくする。

2. 漬物容器に黒砂糖と梅を交互に入れる。

3. キッチンと詰めていき、最後の層には必ず黒砂糖である。

4. 粗塩を加える。手で梅と黒砂糖を軽く押し、梅エキスが滲み出ると、砂糖が溶けしやすくなる。

5. 米酢をかける。ラップをし、日付をつける。

6. 約一週間後に梅エキスはどんどん滲み出る。浅い黄色になり、黒砂糖もどんどん溶ける。このとき、梅と黒砂糖をよく混ぜましょう。

7. 約2週間後に黒砂糖は梅エキスと混ぜ合わせて溶け続け、黒糖シロップに梅を浸す状態になる。梅も柔らかくなり始め黄金色になる。

8. このとき、広口の密閉できる透明なガラス容器に梅と黒糖シロップを入れ、蓋を閉め、日付をつけ、冷蔵庫に入れて保存する。

9. 約2ヵ月後に梅エキスは黒砂糖を完全に溶け、黒糖梅シロップになる。梅酒のように、梅の香りがあふれ出す。でも、製作過程は梅酒のようにそんなに長い時間で待つではないだろう。

Homemade salted green plums
自家製梅干し..........P44

Make it from scratch = it's safer, more hygienic and you'd enjoy it even more.

[Ingredients]
1.2 kg fully ripened green plums
1.2 kg coarse salt
1/2 cup white vinegar

[Method]
1. Rinse the plums and drain well. Use a toothpick to remove the stems.
2. In a sterilized container, put in a layer of coarse salt. Top with a layer plums. Repeat this step until all ingredients are used. Make sure the top-most layer is salt. Pour in the vinegar. Seal the container and label with the production date. Leave the plums in a dry cool place away from the sun for at least 6 months. The salt should dissolve by then and the plums should turn brown and wrinkly.

[Tips]
1. The vinegar actually helps the salt dissolve. You cannot taste the vinegar in the plums.
2. The salted green plums can be used as a condiment in various recipes. Or, you may put them in different drinks. Try salted plum honey drink, salted plum soda, or salted plum with lemon lime soda.

自家製で，，，もっと安全！もっと安心！もっと清潔！

[材料]
完熟梅...1200g
粗塩...1200g
酢...半杯

[作り方]
1. 梅を洗い、水気を取り、竹串で梅のヘタを取り出す。
2. 消毒した容器を使い、まず第一層に粗塩を入れ、次は梅を入れ、そのまま繰り返しして、最後に必ず粗塩を入れる。酢を入れ、蓋を閉め、漬け込んだ日付をつけ、冷暗乾燥と直射日光が当たらない場所に保存し、約6ヶ月で粗塩を完全に溶け、梅は完全に茶色になって、皮が皺になったら出来上がる。

[ポイント]
1. 漬けるときに酢を加える目的は、塩がもっとはやく溶けられ、漬けの味を影響しない。
2. 漬けた梅干しは調味料として、いろんな料理に加えられ、又は、いろんなカクテルを作る。例えば、梅の蜂蜜、梅7−UP、梅ソーダー等。

Steamed swimmer crabs with salted green plums
わたり蟹の梅蒸し.........P46

[Ingredients]

2 swimmer crabs (about 600 g each)
10 homemade salted green plums
3 to 4 tbsps olive oil
finely diced spring onion

[Seasoning]

2 tbsps water
2 tbsps rice wine
3 tbsps tomato paste
1 tbsp plum sauce
1 1/2 tbsps sugar
1/2 tsp light soy sauce

[Method]

1. Dress and rinse the crabs. Chop into pieces. Arrange on a steaming plate.
2. Seed the salted green plums. Crush them and mix well with the seasoning. Heat some olive oil in a pan. Pour in the seasoning and plums. Stir well. Spread the resulting mixture over the crabs. Boil water in a wok or steamer. Steam over high heat for 8 to 10 minutes. Sprinkle spring onion on top. Serve.

[Tips]

1. Though I asked you to seed the salted green plums before use, their seeds actually are quite flavourful. If you don't mind the presentation being a bit less refined, you may actually steam the crabs together with the plum seeds.

[材料]

渡り蟹...2匹(一匹約600g)
自家製梅干し...約10粒
オリーブオイル...3〜4大さじ
ねぎ（小口切り）...適量

[調味料]

水...2大さじ
米酒...2大さじ
ケッチャップ...3大さじ
梅ペースト...1大さじ
砂糖...1 1/2大さじ
醤油...1/2小さじ

[作り方]

1. カニをよく洗い、解体し、胴体を適当なサイズに切り、皿に盛る。
2. 梅干しの種を取り去り、押し潰し、調味料と混ぜ合わせる。オリーブオイルを熱し、調味料を入れ炒め、梅ペーストになり、カニの上にかける。鍋に水が沸騰してから、カニを入れ、強火に約8〜10分間蒸し、取り出してからカニの上にねぎをかけて出来上がる。

[ポイント]

1. 実は自家製梅干しの種が味もよい、気にしなければカニを蒸すときに種と一緒に蒸すのがもっと味がでる。

Braised chicken with young ginger and salted plums
梅、新しょうがと鶏の煮物..........P48

[Ingredients]

1 freshly dressed chicken (about 1.5 kg)
50 g fresh young ginger
100 g sweet vinegar pickled young ginger
rice wine

[Marinade]

1 tbsp dark soy sauce

[Stuffing]

80 g homemade salted green plums (about 10 plums)
3 cloves garlic
40 g crushed rock sugar
1 tbsp ground soybean paste

[Braising stock]

1 tbsp light soy sauce
1 tbsp dark soy sauce
1 clove star anise
1 litre chicken stock

[Method]

1. Rinse the fresh young ginger and crush them gently with the flat side of a knife. Add 1/2 tsp of coarse salt and stir well. Leave them for 30 minutes. Pit the salted plums and set aside the pits for later use. Crush pitted plums, garlic and rock sugar with mortar and pestle. Then add ground soybean sauce. Mix well. This is the stuffing.

2. Rinse the chicken. Wipe dry. Brush marinade all over the chicken. Marinate for 30 minutes. Stir fry the stuffing from step 1 together with the plum pits over low heat until the rock sugar melts. Leave it to cool. Spread the resulting mixture evenly on the inside of the chicken. Then seal the cut with a metal skewer.

3. Heat 1 tbsp of oil in a wok. Stir fry fresh young ginger until fragrant. Sizzle with rice wine. Add the braising stock. Put in the chicken. Cover the lid and bring to the boil over high heat. Turn to low heat and simmer for 15 minutes. Flip the chicken upside down and cook for 15 more minutes until the chicken is thoroughly cooked.

4. Set the chicken and the young ginger aside. Skim off the grease on the surface of the braising stock. Then transfer the stuffing inside the chicken into the stock. Add sweet vinegar pickled ginger. Cook over high heat until the stock reduces to 1 cup. Strain the stock. Thicken with caltrop starch solution to make a light glaze.

5. Chop the chicken into pieces. Pour the thickened glaze over the chicken. Arrange the fresh young ginger around the chicken. Garnish and serve.

[材料]

丸鶏...1羽（1500g)
新しょうが...50g
甘酢しょうが...100g
米酒(調理酒) ...少々

[下味用つけ汁]

老抽 (ラオ　チョウ)...1大さじ

[フィリング]

梅の塩漬け...80g(約10粒)
にんにく...3粒
氷砂糖...40g
磨豉醤 (中国味噌) ...1大さじ

[煮込みソース]

醤油と老抽...各1大さじ
八角...1粒
チキンスープ...約1リットル

[作り方]

1. 新しょうがをよく洗い、押しつぶし、約1/2小さじの粗塩で30分ぐらい下味する。梅干しの種を取っておき(種を残し)、梅干しの身、にんにく、氷砂糖を砕け、磨豉醤と混ぜ合わせておく。

2. 丸鶏をよく洗い、水気を取り、丸鶏の表面に下味用つけ汁をまんべんなく塗り、約30分に下味する。フィリングと梅干しの身を油少々で、氷砂糖を解けるまで炒める。丸鶏のお腹に詰め込んで、チキン針でとじ合わせる。

3. 火をつけ、油1大さじをフライパンに入れ、熱くなったら新しょうがを香ばしく炒め、米酒少々、煮込みソース、丸鶏の順で入れ、蓋をし強火で沸騰してから弱火に15分ほど煮込み、丸鶏の反対側も同様に15分ほど煮込む。

4. 丸鶏と新しょうがを取り出し、残った汁の油分を漉し取る、丸鶏のお腹にフィリングと甘酢しょうがを汁に加え、強火に約コップ一杯の量まで煮る、こし器で漉し分け、とろみをつける。

5. 丸鶏をぶつ切りにして、鶏の上に汁をかけ、漉し分けた新しょうがを鶏の周りに盛り、召し上がる。

Strawberry jam

[Ingredients]

1 kg strawberries
200 g sugar
1/2 tsp salt
juice of 1 lime

[Method]

1. Rinse the strawberries. Wipe dry. Cut off the leaves and stems. Dice finely. Mix strawberries with sugar, salt and lime juice. Cover with cling film. Leave them in the fridge for 1 day, to allow moisture to be drawn out of the strawberries.
2. Transfer the strawberries together with the syrup into a pot. Bring to the boil over low heat. Skim off the bubbles. Leave them to cool. Keep in the fridge for 1 day.
3. Strain the strawberries. Put the syrup into a pot. Boil over high heat. Then turn to low heat and simmer until it thickens (for about 10 minutes). Then put the strawberries into the pot. Mix well. Cook over low heat for 5 minutes. By the time the syrup thickens and bubbles and the strawberries turns glossy, the jam is done. If you like your jam to be runnier, you can turn off the heat at this point. If you'd like it thicker, you may keep on cooking the mixture for a bit longer.
4. Remove from heat. Leave it to cool. Transfer into sealable bottles.

Variations: Regular jam / muscovado jam / whole strawberry jam

[Regular jam]

1. Jam made this way has separate solid fruit embedded in the liquid syrup. The syrup tends to be brightly coloured and clear. The fruit is chunky. If you prefer more conventional jam in which the fruit is mushier, you may boil the strawberries and syrup in step 2 over high heat. Then cook them over low heat until the mixture bubbles and turns thick. This will be the conventional jam.

[Muscovado jam]

1. You may also use muscovado instead of white sugar for a richer flavour and stickier texture.

[Whole strawberry jam]

1. You may use mini strawberries from Japan or Korea for this recipe. Keep the strawberries in whole and follow step 1 to 3. Then store them in bottles. It looks lovely and the whole strawberries can be used in dessert or cakes directly.

[Tips]

1. There will be bubbles and cloudy bits on the surface when you cook the syrup and strawberries. Skimming them off will make the jam look more attractive and less cloudy.
2. I prefer using long-stem strawberries from the US for this recipe, because of their outstanding fragrance, sweetness and flavour. The jam ends up irresistible with its fragrance and gorgeous colour.
3. You may spread the jam on bread, use it in cakes, dribble over yoghurt or ice cream etc. Anyhow, it tastes equally great.

[材料]

いちご...1000g
砂糖...200g
塩...1/2小さじ
ライム汁...1個

[作り方]

1. いちごは洗ってヘタを取り、水気を拭き取り、み
   じん切る。砂糖、塩、ライム汁と混ぜ合わせ、ラ
   ップをする。冷蔵庫に入れ、いちごの汁を出させ
   るため一日置く。
2. 第二日にいちごと出た汁を一緒に鍋に入れ、弱火
   で煮、泡を取り、冷ましてから再びに冷蔵庫に入
   れ一日置く。
3. 第三日にストレーナーでいちごと汁を分け、鍋に
   汁を入れ、強火で沸騰してから弱火にかけ、濃く
   なるまで（約10分間）、いちごを入れ混ぜ、約5
   分間そのまま弱火で煮る。この時に汁は泡が立ち
   始め、いちごもピカピカになり始め、ジャム状に
   なる。基本的には完成した。自分の好みにジャム
   の濃度になるまで火を止める。短時間煮ると、汁
   が多くなり、長時間煮ると汁が濃くなる。
4. ジャムを盛り、冷ましてから容器に入れ、蓋を閉
   め保存してゆっくり召し上がる。

別の風味：伝統ジャム、黒糖ジャム、粒粒ジャム

[伝統ジャム]

1. 上記の方法で作ったジャムは果肉と汁がちゃんと
   分けられ、色が鮮やか、果肉は口触りが良い。も
   し果肉と汁がもっと濃くに混ぜ合わせたくて、伝
   統ジャムのように果肉がジャムの感覚するなら。
   作り方の手順2に、いちごと出た汁を直接に鍋に
   入れ強火で沸騰させてから、弱火に泡が立つ、濃
   くなるまで煮る。伝統ジャムが出来上がる。

[黒糖ジャム]

1. 白い砂糖の代わりに黒砂糖を使う。こうしたら、
   別の風味がある。

[粒粒ジャム]

1. 日本産又は韓国産の小さないちごを手順1，
   2，3，でジャムを作る。完成した後にボトルに
   入れ、見た目が可愛い。ケーキーやデザートな
   どに加えて、何よりも美味い。

[ポイント]

1. 汁が沸騰した頃に泡が浮き上がってきたら、泡
   と滓をすくい取る。作ったジャムは透明感があ
   って、見た目がもっと美しくなる。
2. アメリカ産いちごを選ぶのをお勧めです。そ
   の品種のいちごは香りと味もとても濃厚で、作
   ったジャムはもっと美味しくなる！
3. 出来上がったいちごジャムはパン、ケーキー、
   ヨーグルト、アイスクリームなどと一緒に召し
   上がるとより一層美味しい。

Yoghurt with fruit preserve
ジャム入りヨーグルト..........P53

[Ingredients]

1/2 cup plain yoghurt
homemade fruit preserve

[Method]

1. Scoop some fruit preserve over the plain yoghurt. Mix well before eating.

[Tips]

1. Make sure you refrigerate the fruit preserve after opening the bottle.

[材料]

オリジナルヨーグルト...約１/２カップ
ジャム...適量）

[作り方]

1.召し上がる時に、適量なジャムをヨーグルトに入れ、混ぜてから召し上がる。

[ポイント]

1.開栓後のジャムは必ず冷蔵庫に入れて保存してください。

Candied strawberries in honey
いちごのはちみつシロップ..........P54

[Ingredients]

1.5 kg mini strawberries from Japan or Korea
1 kg honey

[Method]

1. Rinse the strawberries. Cut off the stems. Wipe dry with paper towel. Save in a sealable container. Pour in the honey and seal well. Refrigerate for 1 month until the juice is drawn out of the strawberries and they shrink and turn wrinkly.

[Tips]

1. Throughout the marinating process, stir the mixture once a week. It's because the strawberry juice drawn out of the fruit tends to be less dense than the honey. It will separate and float on top. Stirring the mixture helps mixing the honey and the strawberry juice, so that the strawberries are candied evenly and the honey picks up the strawberry scent and taste better.

2. The strawberries should be candied in whole. Do not cut them into pieces. Otherwise, the honey will turn murky and look less appetizing.

[材料]

日本産或いは韓国産いちご（小粒）...900g
はちみつ...600g

[作り方]

1.いちごを洗い、ヘタを取る、キチンペーパーで水気を取り、容器に入れ、はちみつを入れ蓋を閉め、約一ヶ月で冷蔵庫に置き、いちごの水分が出て、体積を縮むまで出来上がる

[ポイント]

1.放置期間に、いちごから水が出るため、濃度が薄いので、上に浮かび、大体一週間おきに攪拌し、いちご汁がちみつと良く混ぜ合わせさせ、はちみつにいちごの香りと風味が出られる。

2.いちごをカットしなくて丸ごとを漬けて下さい。こうしないと、漬け上がったはちみつの色が濁りになり、雑質が多くて見た目はよくない。

Zesty strawberry wine
情熱のいちごワイン..........P56

[Ingredients]

2.7 kg US long-stem strawberries
2 Thai limes
3 litres rice wine
1.35 kg rock sugar

[Method]

1. Rinse the strawberries and cut off the stems. Drain well. Wipe them completely dry with paper towel. Leave them in a strainer for 1 to 2 hours until they are air-dried. Set aside. Rinse the limes and wipe dry. Cut each lime into halves. Remove the seeds.
2. Put the strawberries, limes, rice wine and rock sugar into a sealable wide-mouth transparent glass container. Seal well. Write down the production date. Leave it in a cool dry place away from the sun. Shake the container once in a while to speed up infusion. After the rock sugar dissolves, leave it still for 6 months. Serve.

[Tips]

1. The tart limes partly balance out the sweetness of strawberries. They also heighten the sugary fruity flavour of the strawberries.
2. You may adjust the amount of rock sugar used according to your own taste.
3. I prefer using US long-stem strawberries for this recipe, because of their outstanding fragrance, taste and sweetness. They are the perfect candidate for making fruit wine.

[材料]

アメリカン産イチゴ...約2.7kg
タイのライム...2個
米酒...3リットル
氷砂糖...1.35kg

[作り方]

1. イチゴを洗い、ヘタを取る、キチンペーパーで水気を取り、ストレーナーに入れ、水気が完全になくなるまで1〜2時間ほど置き、ライムを洗い、水気を取り、半分に切り、種を取り出す。

2. 広口で密閉できるガラス容器にイチゴ、ライム、米酒、氷砂糖を入れ、蓋を閉め、漬け込んだ日付をつけ、冷暗乾燥と直射日光が当たらない場所に保存し、時々容器を振り動かして、氷砂糖を完全に溶けるまで、静止した状態で6ヶ月放置してから召し上がる。

[ポイント]

1. ライムを入れる目的はイチゴの甘さを中和し、イチゴの香りと甘みをもっと強めくなる。
2. 氷砂糖を入れる量は自分の好みで加減する。
3. アメリカン産のイチゴを使用するのはお勧め、その香りと甘みが他の品種より濃く、もっと良い果実酒を作れる。

Hard liquor with high alcohol content efficiently extracts the fragrance of the fruits. In this recipe, Chinese bayberries are soaked in sorghum liquor. The resulting wine has a deep sweet fragrance with a rich fruity taste.

[Ingredients]

1.35 kg fresh Chinese bayberries
1. 35 kg sorghum liquor (55% alcohol)
675 g rock sugar
rice wine (enough to cover the berries)

[Method]

1. Rinse the bayberries and drain well. Soak them in rice wine for 6 to 8 hours. Then drain well and set aside.
2. Put the bayberries, sorghum liquor and rock sugar into a sealable wide-mouth transparent glass container. Seal well. Write down the production date. Leave it in a cool dry place away from the sun. Shake the container once in a while to speed up infusion. After the rock sugar dissolves, leave it still for 6 months. Serve.

[Tips]

1. Chinese bayberries do not have skin on them. Their flesh is moist and hangs like strands on the surface. As the flesh is exposed to the air throughout the growing process, there might be insects or micro-organisms in the flesh. In the wine making process, the fruit has to be soaked in the wine for a prolonged period. It's thus advisable to soak the bayberries in rice wine with over 30% alcohol content for half a day first, so as to sterilize them. The bayberry wine made this way has longer shelf life and is more hygienic for consumption.
2. In fact, even when you serve fresh Chinese bayberries straight as a fruit, you might as well soak them in salted water for 30 minutes first before serving. Firstly, it sanitizes the fruit. Secondly, the bayberries taste sweeter after being soaked in salted water, just like what happens when people sprinkle salt on watermelon.

アルコール度が高いと果実の香味がもっと引き出せる。高粱酒（コウリャン酒）で漬けた果実酒は、香りと味はもっと芳醇に仕上がる。

[材料]

新鮮なヤマモモ...810g
中国の高粱酒（コウリャン酒）...810g（55度）
氷砂糖...675g
米酒...適量（ヤマモモ全体を浸せる量）

[作り方]

1. ヤマモモを洗い、完全に水気を取り、米酒で6〜8時間ヤマモモを浸し、ヤマモモを取り出し、漉し器でこして置く。
2. 広口で密閉できる透明なガラス容器にヤマモモ、氷砂糖、高粱酒を入れ、蓋を閉め、日付をつけ、冷暗乾燥と直射日光が当たらない場所に保存し、たまに容器を揺り動かして、氷砂糖を完全に溶けるまで、静止した状態で約6ヶ月間放置してから召し上がる。

[ポイント]

1. ヤマモモの表面はつぶつぶ粒状突起で水分が多い。昆虫が果実に寄生し、細菌も繁殖しやすい。ヤマモモ酒を作る前に必ず30度以上の米酒でヤマモモを半日に浸して消毒する。こうすると、漬け上がったヤマモモ酒は安全・安心と長時間に保存できる。
2. 普段にヤマモモをそのまま食べでも、塩水で30分ほど浸してから召し上がる。消毒はもちろん、スイカに塩の原理と同じで、ヤマモモにも塩かけると甘くなる。

Guava wine
グアバ酒..........P62

[Ingredients]

800 g local guavas
1 litre Triple-distilled rice wine
250 g rock sugar

[Method]

1. Rinse the guavas. Wipe dry. Cut each into halves.
2. Put the guavas, rice wine and rock sugar into a sealable wide-mouth transparent glass container. Seal well. Write down the production date. Leave it in a cool dry place away from the sun. Shake the container once in a while to speed up infusion. After the rock sugar dissolves, leave it still for 3 months. Serve.

[Tips]

1. I prefer using local pink guavas for this recipe because of their strong fragrance. The guava wine will end up in rich flavour.
2. You have to pick guavas of the right ripeness for the best result. Guavas too unripe are not fragrant enough. Guavas too ripe are mushy and tend to break down after soaked in the rice wine for months. The wine will be cloudy instead of clear.

[材料]

グアバ...８００g
中国の三蒸米酒...１リットル
氷砂糖...２５０g

[作り方]

1. グアバを洗い、水気を取り、半分切る。
2. 広口で密閉できる透明なガラス容器にグアバ、三蒸米酒、氷砂糖を入れ、蓋を閉め、日付をつけ、冷暗乾燥と直射日光が当たらない場所に保存し、時々容器を揺り動かして、氷砂糖を完全に溶けるまで、静止した状態で３ヶ月放置してから出来上がる。

[ポイント]

1. グアバはピンク色の果肉を選び、ピンクグアバとも呼ばれる。芳香がとても強く、漬けたグアバ酒はもっとの香りが漂って来る。
2. 成熟程度は丁度いいグアバを選らんだら、美味しいグアバ酒を作られる。熟していないと香りが薄く、熟しすぎると果肉が柔らかすぎ、長時間に漬けると果肉が分解しやすく、果実酒が濁ってしまう。

Aged osmanthus Chinese bayberry wine
ヤマモモ桂花陳酒..........P66

[Ingredients]

1.35 kg fresh Chinese bayberries
3 bottles aged osmanthus wine (750 ml each)
rice wine (enough to cover the berries)

[Method]

1. Rinse the bayberries and drain well. Soak them in rice wine for 6 to 8 hours. Then drain well and set aside.
2. Put the bayberries and aged osmanthus wine into a sealable wide-mouth transparent glass container. Seal well. Write down the production date. Leave it in a cool dry place away from the sun for 6 months. Serve.

[Tips]

1. Please refer to "Homemade Chinese bayberry wine" on p231.

[材料]

新鮮なヤマモモ...1.35kg
桂花陳酒...3ボトル（1ボトルは750ml）
米酒...適量（ヤマモモが浸る量）

[作り方]

1. ヤマモモを洗い、完全に水気を取り、米酒で6～8時間ヤマモモを浸し、ヤマモモを取り出しから漉し器でこして置く。
2. 広口で密閉できる透明なガラス容器にヤマモモと桂花陳酒を入れ、蓋を閉め、日付をつけ、冷暗乾燥と直射日光が当たらない場所に保存し、静止した状態で約6ヶ月間放置してから召し上がる。

[ポイント]

1. 231ペッジ（自家製ヤマモモ酒）に参考する。

Apricot rum

[Ingredients]

800 g dried apricots
1 litre white rum

[Method]

1. In a sealable bottle, put in the dried apricots and rum. Seal well. Leave them in a cool dry place away from the sun for 6 months. Serve.

Raisin Vodka

[Ingredients]

500 g raisins
750 ml currant vodka (1 bottle)

[Method]

1. In a sealable bottle, put in the raisins and vodka. Seal well. Leave them in a cool dry place away from the sun for 6 months. Serve.

Longan Rum

[Ingredients]

560 g dried longans (shelled and seeded)
1.5 litre gold rum (2 bottles)

[Method]

1. In a sealable bottle, put in the longans and gold rum. Seal well. Leave them in a cool dry place away from the sun for 6 months. Serve.

[Tips]

1. Try to get individually packed dried apricots and raisins for instant consumption. They are more hygienic than those available in bulk. You don't need to rinse the dried fruits before steeping them in wine. In case there is any moisture on the fruit, the wine will turn stale easily.
2. Try to get dried longans that have not been bleached by sulphur dioxide. This way, the fruit wine will not be contaminated by additives and you'd have the peace of mind.

ドライアプリコット・ラム酒

[材料]

ドライアプリコット...800g
ホワイトラム...1リットル

[作り方]

1. 容器にホワイトラム酒を入れ、ドライアプリコットを加え、蓋を閉め、冷暗乾燥と直射日光が当たらない場所に保存し、6ヶ月漬けたら出来上がる。

レーズンウォッカ

[材料]

レーズン...500g
カシスウォッカ（白）...750ml（1ボトル）

[作り方]

1. 容器にウォッカを入れ、レーズンを入れ、蓋を閉め、冷暗乾燥と直射日光が当たらない場所に保存し、6ヶ月漬けたら出来上がる。

干しロンガン・ラム酒

[材料]

干しロンガン...560g
ゴールドラム...1.5リットル（2ボトル）

[作り方]

1. 容器にラム酒を入れ、干しロンガンを入れ、蓋を閉め、冷暗乾燥と直射日光が当たらない場所に保存し、6ヶ月漬けたら出来上がる。

[ポイント]

1. 干しロンガンとレーズンは個別包装したものを選んだほうが安全である。水気があると腐りやすいので干しロンガンとレーズンは洗う必要がない。
2. 硫黄と漂白処理をしていない干しロンガンを買う。こうして、オリジナルな味でドライフルーツ酒が漬けられ、無添加で健康に飲みましょう。

Japanese peach wine
日本のもも酒.........P68

[Ingredients]

1.6 kg Japanese white peaches (about 6 peaches)
1.8 litres rice wine
600 g rock sugar

[Method]

1. Rinse the peaches and wipe dry. Cut each into halves.
2. Put peaches, rice wine and rock sugar into a sealable wide-mouth glass container. Seal well and write down the production date. Leave it in a cool dry place away from the sun. Shake the container once in a while to speed up infusion. After the rock sugar dissolves, leave it still for 6 months. Serve.

[材料]

日本産もも...1600g（約6個）
米酒...1.8リットル
氷砂糖...600g

[作り方]

1. ももを洗い、水気を取り、半分に切る。
2. 広口で密閉できる透明なガラス容器にもも、米酒と氷砂糖を入れ、蓋を閉め、日付をつけ、冷暗乾燥と直射日光が当たらない場所に保存し、時々容器を揺り動かして、氷砂糖を完全に溶けるまで、静止した状態で約6ヶ月放置すると出来上がる。

Homemade raspberry wine
自家製ラズベリー酒..........P76

[Ingredients]

320 g fresh raspberries
600 ml Guangxi rice wine
200 g rock sugar

[Method]

1. Rinse the raspberries and drain well. Wipe them completely dry with paper towel. Leave them in a strainer for 1 to 2 hours until they are air-dried.
2. Put the raspberries, rice wine and rock sugar into a sealable wide-mouth transparent glass container. Seal well. Write down the production date. Leave it in a cool dry place away from the sun. Shake the container once in a while to speed up infusion. After the rock sugar dissolves, leave it still for 3 months. Serve.

[Tips]

1. Before you make wine with the raspberries, make sure they are completely dry. Unboiled water carries impurities and micro-organisms. Fruit wine with unboiled water in it turns stale very easily and has a very short shelf life.

[材料]

新鮮なラズベリー...320g
中国の広西米酒...600ml
氷砂糖...200g

[作り方]

1. ラズベリーを洗って水切り、キッチンペーパーで水気を拭き取り、水気を完全に取るようにざるに1〜2時間放置する。
2. 広口で密閉できる透明なガラス容器にラズベリー、米酒、氷砂糖を入れ、蓋を閉め、日付をつけ、冷暗乾燥と直射日光が当たらない場所に保存し、たまに容器を揺り動かして、氷砂糖を完全に溶けるまで、静止した状態で約3ヶ月間放置してから召し上がる。

[ポイント]

1. ラズベリーを必ず水気を完全に取る。生水の中には雑質や細菌があり、果実酒を腐らせてしまい、長時間に保存できなくなる。

[Ingredients]

1 kg black mulberries
1.2 litre dark rum
150 g rock sugar

[Method]

1. Rinse the black mulberries. Drain and wipe dry thoroughly with paper towel. Leave them in a strainer for 1 to 2 hours to make sure they are air-dried completely.
2. Prepare a sealable, wide-mouth transparent glass container. Put in the black mulberries, dark rum and rock sugar. Seal the container and write down the date of preparation. Leave it in a cool dry place in the shade away from the sun. Shake the container once in a while until the rock sugar dissolves. Leave it still for 3 months. Serve.

[Tips]

1. Pick black mulberries that are dark purple in colour and plump in shape.
2. There are many creases on mulberries that may trap moisture. Thus, make sure you dry them thoroughly before use. Otherwise, the bacteria and impurities in the uncooked water would make the rum stale easily and significantly reduce its shelf life.

[材料]

黒桑の実...1000g
黒ラム酒...1.2リットル
氷砂糖...150g

[作り方]

1. 黒桑の実を洗い、水気を取り、キチンペーパーで水気を拭き完全に乾くまで1～2時間に置く。
2. 透明な密閉ガラス容器に黒桑の実、黒ラム酒、氷砂糖を入れ、蓋を閉め、日付をつけ、冷暗乾燥と直射日光が当たらない場所に保存し、時々容器を揺り動かして、氷砂糖を完全に溶けるまで、静止した状態で約3ヶ月放置してから召し上がる。

[ポイント]

1. 実の表面がふくらみ、深い紫色の黒桑の実を選んでください。
2. 黒桑の実の表面はつぶつぶの突起で小さいな隙間が多く、水気が留まりやすいので、水気をよく取ってください。生水は雑質と細菌が多いから、果実酒が腐りやすく、長時間に保存できなくなる。

Kyoho grape wine
日本巨峰酒..........P78

[Ingredients]

1 kg Japanese Kyoho grapes
1 litre Triple-distilled Chinese rice wine
120 g rock sugar

[Method]

1. Rinse the grapes and drain well. Wipe them dry thoroughly with paper towel.
2. Prepare a sealable, wide-mouth transparent glass container. Put in the grapes, rice wine and rock sugar. Seal well and label it with the date of preparation. Leave it in a cool and dry place in the shade away from direct sunlight. Shake the container once in a while until the rock sugar dissolves completely. Leave it still for 3 months. Serve.

[材料]

巨峰...1000g
中国の三蒸米酒...1000ml
氷砂糖...120g

[作り方]

1. 巨峰を洗い、水気を取り、キッチンペーパーで水気を良く拭き取る。
2. 透明な密閉ガラス容器に巨峰、三蒸米酒、氷砂糖を入れ、蓋をし、日付をつけ、乾燥した直射日光が当たらない冷暗所で保存し、たまに容器を揺らして、砂糖を完全に溶かし、静止した状態で約3ヶ月間に置いてから召し上がる。

Japanese Ourin apple wine
日本の王林りんご酒..........P80

[Ingredients]

1.5 kg Japanese Ourin apples (about 7 apples)
2.8 litres Triple-distilled rice wine
380 g rock sugar

[Method]

1. Rinse the apples. Wipe them completely dry with paper towel and leave them to air dry. Prick 5 to 6 holes on each apple with a bamboo skewer for easy infusion of the apple juice.
2. Put the apples, rice wine and rock sugar into a sealable wide-mouth transparent glass container. Seal well. Write down the production date. Leave it in a cool dry place away from the sun. Shake the container once in a while to speed up infusion. After the rock sugar dissolves, leave it still for 6 months. Serve.

[Tips]

1. Please refer to "Homemade apple vinegar" on p.261.

[材料]

王林りんご...1500g（約7個）
中国三蒸米酒...2.8リットル
氷砂糖...380g

[作り方]

1. りんごを洗い、キッチンペーパーで水気を拭き取り、エキスを出やすくするため、竹串で5〜6個穴を刺す。
2. 広口で密閉できる透明なガラス容器にりんご、三蒸米酒、氷砂糖を入れ、蓋を閉め、日付をつけ、冷暗乾燥と直射日光が当たらない場所に保存し、たまに容器を揺り動かして、氷砂糖を完全に溶けるまで、静止した状態で約6ヶ月を放置してから召し上がる。

[ポイント]

1. 261ページ自家製りんご酢に参考する。

Homemade golden plum wine
自家製すもも酒..........P86

[Ingredients]

2.2 kg Chinese golden plums
1.8 kg rice wine
1.2 kg rock sugar

[Method]

1. Rinse the plums. Remove the stems. Drain well and wipe them completely dry with paper towel. Prick a few holes on each plum with a bamboo skewer for easy infusion of the plum juice.
2. Put the plums, rice wine and rock sugar into a sealable wide-mouth transparent glass container. Seal well. Write down the production date. Leave it in a cool dry place away from the sun. Shake the container once in a while to speed up infusion. After the rock sugar dissolves, leave it still for 6 months. Serve.

[Tips]

1. This wine is sour and sweet in taste, making it a great aperitif to be served before meals. It sort of tastes like ume wine, but milder in taste and lighter in fragrance.
2. Besides Chinese yellow plums, you may actually use greengage, black plums or prunes instead. They make different wines which taste equally great.

[材料]

黄色すもも...2200g
米酒　1800g
氷砂糖...1200g

[作り方]

1. すももを洗い、ヘタをとり、キッチンペーパーで水気を拭き取り、すももに幾つの穴を刺し、汁を出やすくさせる。
2. 広口で密閉できる透明なガラス容器にすもも、米酒、氷砂糖を入れ、蓋を閉め、日付をつけ、冷暗乾燥と直射日光が当たらない場所に保存し、時々容器を揺り動かして、氷砂糖を完全に溶けるまで、静止した状態で6ヶ月放置してから出来上がる。

[ポイント]

1. すももで漬けた果実酒は甘みと酸味が調和した味わい。食欲を増進し、食前酒にふさわしい。梅酒の似た風味ですが、味がマイルドで香りが自然に立ち上がる。
2. すももの他に、青プラム、黒プラム、プルーンなどで漬けた果実酒は効果も良く、別の風味がある。

Cherry wine
サクランボ酒...........P84

[Ingredients]

1 kg US dark cherries
1.2 litre Triple-distilled rice wine
500 g rock sugar

[Method]

1. Rinse the cherries. Remove the stems. Drain well. Wipe them completely dry with paper towel.
2. Put the cherries, rice wine and rock sugar into a sealable wide-mouth transparent glass container. Seal well. Write down the production date. Leave it in a cool dry place away from the sun. Shake the container once in a while to speed up infusion. After the rock sugar dissolves, leave it still for 3 months. Serve.

[Tips]

1. I prefer dark cherries from the US because of its strong taste and sweetness. They make the best ingredient for making fruit wine.

[材料]

アメリカ産サクランボ...1000g
中国の三蒸米酒...1.2リットル
氷砂糖...500g

[作り方]

1. サクランボを洗い、ヘタを取り、水を切り、キッチンペーパーで水気を完全に拭き取る。
2. 広口で密閉できる透明なガラス容器にサクランボ、三蒸米酒、氷砂糖を入れ、蓋を閉め、日付をつけ、冷暗乾燥と直射日光が当たらない場所に保存し、たまに容器を揺り動かして、氷砂糖を完全に溶けるまで、静止した状態で約3ヶ月間放置してから召し上がる。

[ポイント]

1. アメリカ産の濃い赤紫色のサクランボを選んでおすすめです。味が甘くて濃厚で、果実酒を漬けるのが一番良い。

Blackberry rum
ブラックベリーラム酒...........P88

[Ingredients]

480 g fresh blackberries
550 ml dark rum
180 g rock sugar

[Method]

1. Rinse the blackberries and drain well. Wipe them completely dry with paper towel. Leave them in a strainer for 1 to 2 hours until they are air-dried.
2. Put the blackberries, rum and rock sugar into a sealable wide-mouth transparent glass container. Seal well. Write down the production date. Leave it in a cool dry place away from the sun. Shake the container once in a while to speed up infusion. After the rock sugar dissolves, leave it still for 3 months. Serve.

[Tips]

1. Before you make wine with the blackberries, make sure they are completely dry. Unboiled water carries impurities and micro-organisms. Fruit wine with unboiled water in it turns stale very easily and has a very short shelf life.

[材料]

新鮮なブラックベリー...480g
ダークラム...550g
氷砂糖...180g

[作り方]

1. ブラックベリーを洗い、水気を取り、キチンペーパーで拭き取り、ザルに入れて1〜2時間水切りにする
2. 広口で密閉できる透明なガラス容器にブラックベリー、ダークラム、氷砂糖を入れ、蓋を閉め、日付をつけ、冷暗乾燥と直射日光が当たらない場所に保存し、時々容器を揺り動かして、氷砂糖を完全に溶けるまで、静止した状態で3ヶ月放置してから出来上がる。

[ポイント]

1. 洗ったブラックベリーは水分を完全に切るのを確認し、生水は雑質と細菌が含まれ、果実酒が腐りやすく、長期保存はできなくなる。

[Ingredients]

600 g fresh lychees (preferably "Nuo Mi Ci" variety)
1 bottle aged osmanthus wine (750 ml)

[Method]

1. Rinse the lychees. Drain well. Shell and seed them. Set the flesh aside.
2. Put the lychees and osmanthus wine into a sealable wide-mouth transparent glass container. Seal well. Write down the production date. Leave it in a cool dry place away from the sun for 3 months. Serve.

[Tips]

1. Try not to bruise the lychees too much when you shell and seed them. Otherwise, there will be too much juice mixed directly with the wine, making it cloudy instead of clear. When you make fruit wine, you should allow time for the alcohol to extract the fruity flavour from the fruit. If you mix fruit juice with wine, it tastes more like a cocktail.
2. There are over 40 varieties of lychees and each of them tastes slightly different. As a result, wine made with different varieties of lychees also taste different. From my personal experience, lychee wine made with the "Nuo Mi Ci" variety (the most popular and highly prized variety in China) tastes the sweetest and most aromatic. You may also experiment with the proportion between wine and lychees so as to fine tune the strength of lychee taste according to your preference.

[材料]

新鮮なライチ(糯米糍タイプ)...600g
桂花陳酒...1本（1本は750ml）

[作り方]

1. ライチを洗い、水気を取り、皮を剥き、種を取り除く。
2. 広口で密閉できる透明なガラス容器にライチと桂花陳酒を入れ、蓋を閉め、日付をつけ、冷暗乾燥と直射日光が当たらない場所に保存し、約３ヶ月でそのまま放置してから召し上がる。

[ポイント]

1. ライチの果肉を取るとき、果肉を丁寧に取り出してください。果肉を潰したら汁が多く出て酒と混ざると、漬けた酒が濁ってしまう。果実酒を漬けるには十分の時間で果実と酒を混合させ、十分の時間で果実のエキスを吸収させる。
2. ライチの品種によっては甘さと酸味の差があり、漬けた果実酒の味も違いがある。僕の経験により糯米糍（ノーマイチ）で漬けたライチ酒は一番美味いです。皆さんも自分の好みにあったライチ酒の濃淡度を桂花酒とライチの比例を加減する。

Passionfruit pineapple jam
パッションフルーツとパイナップルジャム..........P92

[Ingredients]

1 kg pineapple (peeled, cored)
300 g passionfruit pulp (from about 15
passionfruits)
juice of 1 lime
300 g sugar
1/2 tsp salt
300 ml water

[Method]

1. Dice the pineapple finely. Add passionfruit pulp, sugar, salt and lime juice. Mix well. Cover with cling film. Refrigerate for 1 day until water is drawn out of the pineapple.
2. Transfer the pineapple and passionfruit pulp together with the syrup into a pot. Bring to the boil over low heat. Skim off the foam on the surface. Leave it to cool. Refrigerate for 1 more day.
3. Set aside the pineapple. Pour the passionfruit pulp and syrup into a pot. Add 300 ml of water. Bring to the boil over high heat. Then turn to low heat and simmer until the syrup thickens. Put the pineapple back in. Stir well. Cook over low heat for about 5 minutes until the syrup turns bubbly and thick, while the pineapple turns glossy. The jam is basically done at this stage if you like your jam to be thinner and runnier. Or, you may keep on cooking it longer if you prefer a stiffer jam.
4. Transfer the jam mixture into sterilized bottles. Seal well for storage.

[Tips]

1. It's advisable to pick ripe pineapple for this recipe. This way, the jam will be softer in texture with a stronger pineapple flavour.
2. When you cook the fruit in the syrup, some foam and cloudy bits will float on the surface. You should skim them off, so that the jam will end up clear and has a gem-like quality.
3. In the cooking process, make sure you stir the jam mixture from time to time because the sugar in the jam burns easily.

[材料]

パイナップル（果肉のみ）...1000g
パッションフルーツ(果肉のみ)...300g（約15コ）
ライムジュース...1コ
砂糖...300g
塩...1/2小さじ
水...300ml

[作り方]

1. パイナップルを小口切り、パッションフルーツ、ライムジュース、砂糖、塩を入れ混ぜ、ラップをし、冷蔵庫に入れ、水を出させるため一日置く。
2. 第二日に果肉と出た水を鍋に入れ、弱火で沸騰してから、泡を取り。盛って冷ましてから冷蔵庫に入れ一日置く。
3. 第三日にストレーナーで果肉と汁を分け、汁を鍋に入れ、300ml水を加え、強火で沸騰してから弱火に汁が濃くなるまで煮、果肉を入れ混ぜ、弱火で5分間に続けて煮る。泡が立ち始めるとき、果肉もピカピカになり始め、ジャム状になる。この時は基本的に出来上がる。自分の好みにジャムの濃度になるまで火を止める。短時間煮ると、汁が多くなり、長時間煮ると汁が濃くなる。
4. ジャムを盛り、熱いうちに消毒した容器に入れ、蓋をしてゆっくり召し上がる。

[ポイント]

1. 熟したパイナップルを選んだ方がよい、作ったジャムが柔らかくて、香りが濃くなる。
2. ジャムを煮る間に泡が浮き上がると、泡をすくい取ってください。こうすると、ジャムが透明感があってピカピカ光る。見た目がもっと美しいである。
3. 糖分が高いのでジャムが焦げないように時々かき混ぜながら煮る。

Rose peach jam
モモのバラ入りジャム..........P95

Peach is juicy, sweet and fragrant. When made into a jam, peach has a sophisticated palate. The rose in the jam adds a floral aroma and lends a tempting purplish red hue. This jam is the indispensable breakfast accompaniment in the summer.

[Ingredients]

1 kg Japanese white peaches (about 8 peaches, skinned and seeded)
200 g sugar
1/2 tsp salt
juice of 1 lime
8 dried rose buds (broken into petals)

[Method]

1. Dice the peaches finely. Add sugar, salt and lime juice. Mix well. Cover with cling film. Refrigerate for 1 day until water is drawn out of the peaches.
2. Transfer the peaches together with the syrup into a pot. Bring to the boil over low heat. Skim off the foam on the surface. Leave it to cool. Refrigerate for 1 more day.
3. Transfer the peaches and the syrup into a pot again. Add rose petals. Bring to the boil over high heat. Then turn to low heat and simmer until the syrup thickens and bubbles while the peaches turn glossy. The jam is basically done at this stage if you like your jam to be thinner and runnier. Or, you may keep on cooking it longer if you prefer a stiffer jam.
4. Transfer the jam mixture into sterilized bottles for storage.

[Tips]

1. I prefer Japanese white peaches for their strong flavour. I also prefer them ripe. This way, the jam will have strong peach flavour and natural sweetness, with a satiny mouthfeel.

モモと言えば、汁が多く香りが濃いと思われ、モモで作ったジャムはさっぱりとした甘みに上品な香りがある。更に、バラのほのかな香りに神秘的な紫紅色を加える。このジャムはきっと夏の朝食にぴったり。

[材料]

日本産モモ... 1000g（果肉のみ　約8個）
砂糖...200g
塩...1/2小さじ
ライム汁...約1コ
乾燥バラのつぼみ...約8コ（花びらを散らす）

[作り方]

1. モモを細かく粒を切り、砂糖、塩とライム汁を混ぜ合わせ、ラップをし冷蔵庫に入れ、水を出させるため一日置く。
2. 第二日にモモと出た水を一緒に鍋に入れ、弱火で煮、泡が出てきたらすくい取る。モモと出た水を容器に盛り、冷ましてから冷蔵庫に入れて一日置く。
3. 第三日にモモと水分を鍋に入れ、バラの花びらを加え、強火で沸騰してから弱火にかけ、汁が濃くなり、泡が立ち始めるとき、モモがピカピカになり始め、ジャム状になる。この時には基本的に出来上がる。自分の好みにジャムの濃度になるまで火を止める。短時間煮ると、汁が多くなり、長時間煮ると汁が濃くなる。
4. ジャムを盛り、熱いうちに消毒した容器に入れて、ゆっくり召し上がる。

[ポイント]

1. 日本産の完熟モモを選んでください。こうしたら、モモジャムの香りと甘みがあって、他にはない滑らかな食感が楽しめる。

Honey kiwi jam
はちみつキウイジャム..........P98

[Ingredients]

1 kg kiwis (about 10 kiwis)
100 g sugar
juice of 1 lime
1 tsp salt
120 g honey
300 ml water

[Method]

1. Rinse the kiwis and peel them. Dice finely. Add sugar, lime juice and salt. Mix well. Cover in cling film and leave them in the fridge for 1 day until moisture is drawn out of them.

2. Transfer the kiwi flesh together with the syrup into a pot. Bring to the boil over low heat. Skim off the foam on the surface. Leave it to cool. Refrigerate for 1 more day.

3. Strain the kiwi syrup into a pot. Set aside the flesh for later use. Add honey and water to the syrup. Bring to the boil over high heat. Then turn to low heat and simmer until the syrup thickens. Put the kiwi flesh back in. Stir well. Keep on simmering over low heat for 5 more minutes until the syrup turns bubbly and the kiwi turns glossy. The jam is basically done at this stage if you like your jam to be thinner and runnier. Or, you may keep on cooking it longer if you prefer a stiffer jam.

4. Transfer the jam mixture into sterilized bottles while still hot. Seal them well for storage.

[Tips]

1. You may further explore the nuance in palate by using different kinds of honey, such as multifloral honey, winter honey or Longan honey. The jam will end up tasting slightly different.

2. In the process of cooking the jam mixture, foamy substance appears on the surface. Make sure you skim it off because this is the trick to crystal clear jam that looks appetizing. Besides, make sure you stir the jam from time to time when you cook it. Otherwise, the jam might get burnt.

[材料]

キウイ...600g（約10コ）
砂糖...100g
ライム汁...1コ
塩...1小さじ
ハチミツ...120g
水...300ml

[作り方]

1. キウイを洗い、皮を剥き、小口切り、砂糖、ライム汁、塩と混ぜ合わせ、ラップをし、冷蔵庫に入れ、水を出させるため一日置く。

2. 第二日に果肉と出た水を鍋に入れ、弱火で沸騰し、泡を取り、果肉と汁を盛り、冷ましてから冷蔵庫に入れ一日置く。

3. 第三日にストレーナーで果肉を取り出し、汁を鍋に入れ、ハチミツと水を加え、強火で沸騰してから、弱火に汁が濃くなるまで煮、果肉を入れ、弱火で5分間に煮、泡が立ち始め、果肉もピカピカになり始め、ジャムの状態になり、この時は基本的に出来上がる。自分の好みにジャムの濃度になるまで火を止める。短時間煮ると、汁が多くなり、長時間煮ると汁が濃くなる。

4. ジャムを盛り、熱いうち、消毒した容器に入れ、蓋をしてゆっくり召し上がる。

[ポイント]

1. 各種のハチミツを使用でき、例えば百花蜜、冬のハチミツ、ロンガンのハチミツなど、別の風味がある。

2. ジャムを煮る間に泡が浮き上がると、泡をすくい取ってください。こうすると、ジャムが透明感があってピカピカ光る。ところが、ジャムが焦げないように時々かき混ぜながら煮る。

Vanilla mango jam
バニラとマンゴーのジャム..........P100

[Ingredients]

1 kg Filipino mangoes (weight after peeled and pitted)
200 g sugar
juice of 1 lime
1/2 tsp salt
2 vanilla pods

[Method]

1. Dice the mangoes finely. Add sugar, lime juice and salt. Mix well. Cover with cling film and leave them in a fridge. Leave them for 1 day until moisture is drawn out of the mangoes.

2. Put the mangoes together with the syrup into a pot. Bring to the boil over low heat. Skim off the foam on the surface. Leave the mixture to cool. Keep in the fridge for one more day.

3. Transfer the mangoes with the syrup into a pot. Cut open the vanilla pods and scrape off the black seeds inside. Add the vanilla seeds to the mangoes. Bring to the boil over high heat. Then turn to low heat. Simmer until the syrup thickens and turns foamy. The mango should look glossy and jam-like. The jam is basically done at this stage if you like your jam to be thinner and runnier. Or, you may keep on cooking it longer if you prefer a stiffer jam.

4. Transfer the jam mixture into sterilized bottles while still hot. Seal them well for storage.

[Tips]

1. Pick fully ripened mangoes for this recipe for their irresistible sweetness and flavours.

2. You may use other mangoes instead of Filipino ones for a variation of taste.

[材料]

フィリピンマンゴー...1000g（果肉のみ）
砂糖...200g
ライム汁...1コ
塩...1/2小さじ
バニラ棒...2本

[作り方]

1. マンゴー果肉を粗みじん切り、砂糖、ライム汁、塩と混ぜ合わせ、ラップをし、冷蔵庫に入れ、水を出させるため一日置く。

2. 第二日に鍋にマンゴーと出た汁を入れ、弱火で煮立ってきて浮き上がったアクをすくい取る。盛って冷ましてから冷蔵庫に入れ一日置く。

3. 第三日に鍋にマンゴーと汁を入れ、バニラ棒を縦半分に切り、中身の種をしっかり取り出し、鍋にマンゴーと入れ混ぜ、強火で沸騰してから、弱火にかけ、汁が濃くなり、泡が立ち始める。この時にはマンゴーがピカピカになり始め、ジャム状になり、基本的には完成した。自分の好みにジャムの濃度になるまで火を止める。短時間煮ると、汁が多くなり、長時間煮ると汁が濃くなる。

4. ジャムを盛り、熱いうちに消毒した容器に入れ、蓋をしてゆっくり召し上がる。

[ポイント]

1. 完熟したマンゴーを使うと、ジャムがもっと甘く香ばしく、滑らかになる。

2. 他のマンゴー品種でジャムを作るには別の風味がある。

Star-fruit muscovado jam
黒糖スターフルーツジャム..........P102

I love the natural caramelly sweetness of muscovado, which makes it perfect for different jams and preserves. It also adds a unique flavour quite different from that of conventional jams.

[Ingredients]

1 kg fully ripened star-fruits
200 g muscovado sugar
juice of 1 lime
1/2 tsp salt

[Method]

1. Trim off the ridges on the star-fruits. Core and seed them. Dice finely. Add muscovado sugar, lime juice and salt. Mix well. Seal with cling film. Refrigerate for 1 day until moisture is drawn out of the star-fruits.

2. Transfer the star-fruits together with the syrup into a pot. Bring to the boil over low heat. Skim off the foam on the surface. Leave them to cool. Refrigerate for 1 more day.

3. Strain the star-fruit syrup into a pot. Set aside the flesh for later use. Bring the syrup to the boil over high heat. Then turn to low heat and simmer until the syrup thickens. Put the star-fruit flesh back in. Stir well. Keep on simmering over low heat for 5 more minutes until the syrup turns bubbly and the star-fruit turns glossy. The jam is basically done at this stage if you like your jam to be thinner and runnier. Or, you may keep on cooking it longer if you prefer a stiffer jam.

4. Transfer the jam mixture into sterilized bottles while still hot. Seal them well for storage.

[Tips]

1. You may also use white sugar instead of muscovado. The method and amount remain the same.

2. Pick fully ripened star-fruits for this recipe because they are juicier and have a stronger flavour.

僕は黒糖の焦げた香りと甘み、その自然風味がとても好き。ジャムにするにはふさわしい。そして、伝統的なジャムと違った独特の味わいを与える。

[材料]

完熟スターフルーツ...600g
黒砂糖...200g
ライム汁...1コ
塩...1/2小さじ

[作り方]

1. スターフルーツの角をそぎ落とし、芯を取り去り、種を取り、小口切り、黒砂糖、ライム汁、塩と混ぜ合わせ、ラップをし、冷蔵庫に入れ、水を出させるため一日置く。

2. 第二日に鍋にスターフルーツと出た水を入れ、弱火で沸騰してから、泡を取り。スターフルーツと汁を盛り、冷ましてから冷蔵庫に入れ一日置く。

3. 第三日にストレーナーでスターフルーツを取り出し、汁を鍋に入れ、強火で沸騰してから弱火に汁が濃くなるまで煮、再びにスターフルーツを入れ混ぜ、弱火で5分間に続けて煮る。泡が立ち始めるとき、スターフルーツもピカピカになり始め、ジャム状になる。この時は基本的に出来上がる。自分の好みにジャムの濃度になるまで火を止める。短時間煮ると、汁が多くなり、長時間煮ると汁が濃くなる。

4. ジャムを盛り、熱いうち、消毒した容器に入れ、蓋をしてゆっくり召し上がる。

[ポイント]

1. 黒砂糖の代わりに白砂糖を使用できる。作り方と分量は黒砂糖と同じ（上記）。

2. 熟したスターフルーツを選んだ方がよい、汁は多くて、作ったジャムも味がもっと濃い。

Aged osmanthus lychee wine jam
ライチの桂花陳酒入りジャム........P104

[Ingredients]

1 kg fresh lychees (shelled and seeded)
1 tsp salt
200 g sugar
juice of 1 lime
650 ml aged osmanthus wine

[Method]

1. Dice the lychees finely. Add sugar, salt and lime juice. Mix well. Cover with cling film. Refrigerate overnight until water is drawn out of the lychees.
2. Transfer the lychees together with the syrup into a pot. Add aged osmanthus wine. Bring to the boil over high heat. Then turn to low heat and simmer until the syrup thickens and bubbles while the lychees turn glossy. The jam is basically done at this stage if you prefer the jam to be thinner and runnier. Or, you may keep on cooking it longer if you prefer a stiffer jam.
3. Transfer the jam mixture into sterilized bottles. Seal well for storage.

[材料]

新鮮なライチ果肉...1000g（果肉のみ）
塩...1小さじ
砂糖...200g
ラ·イム汁...1コ
桂花陳酒...650ml

[作り方]

1. ライチ果肉を粒に切り、塩、砂糖とライム汁を入れ混ぜ、ラップをし冷蔵庫に入れ、水を出させるため一日に置く。
2. 翌日にライチ果肉と出た水を鍋に入れ、桂花陳酒を加え、強火で沸騰してから、弱火にかけ、汁が濃くなり、泡が立つまで煮詰める。この時にライチと汁がピカピカになり始め、ジャム状になり、基本的には完成した。自分の好みにジャムの濃度になるまで火を止める。短時間煮ると、汁が多くなり、長時間煮ると汁が濃くなる。
3. ジャムを盛り、熱いうちに消毒した容器に入れ、蓋を閉め保存してゆっくり召し上がる。

Kyoho grape jam
プレミアム巨峰ジャム..........P106

[Ingredients]

1 kg Kyoho grapes
1/2 tsp salt
100 g sugar
juice of 1 lime

[Method]

1. Rinse the grapes and drain well. Peel and seed them. Cut each grape into three pieces or quarters. Add sugar, salt and lime juice. Mix well. Cover with cling film. Refrigerate for 1 day until water is drawn out of the grapes.
2. Transfer the grapes together with the syrup into a pot. Bring to the boil over high heat. Then turn to low heat and simmer until the syrup thickens and bubbles while the grapes turn glossy. The jam is basically done at this stage if you like your jam to be thinner and runnier. Or, you may keep on cooking it longer if you prefer a stiffer jam.
3. Transfer the jam mixture into sterilized bottles. Seal well for storage.

[材料]

巨峰...1000g
塩...1/2小さじ
砂糖...100g
ラ·イム汁...1コ

[作り方]

1. 巨峰を洗い、水気を取り、皮を剥き、種を取り、各粒を3〜4分程度に切る。塩、砂糖、ライム汁と混ぜ合わせ、ラップをし、冷蔵庫に入れ、一日置いて水を出させる。
2. 翌日に鍋に巨峰と出た水を一緒に入れ、強火で沸騰してから弱火にかけ、汁が濃くなり、泡が立ち始める。この時に巨峰と汁がピカピカになり始め、ジャム状になり、基本的には完成した。自分の好みにジャムの濃度になるまで火を止める。短時間煮ると、汁が多くなり、長時間煮ると汁が濃くなる。
3. ジャムを盛り、熱いうちに消毒した容器に入れ、蓋をしてゆっくり召し上がる。

Dried plum and pineapple jam
干し梅とパイナップルジャム.........P108

[Ingredients]

1 kg pineapple flesh
50 g dried preserved plums (about 10 plums)
juice of 1 lime
250 g sugar
3 cups water

[Method]

1. Dice the pineapple. Pit the dried plums and finely chop the flesh. Add dried plums, lime juice and sugar to the diced pineapple. Cover with cling film and refrigerate for 1 day until moisture is drawn out of the pineapple.

2. Pour the pineapple together with the syrup into a pot. Bring to the boil over low heat. Skim off the foam on top. Leave it to cool. Refrigerate for 1 more day.

3. Strain the pineapple syrup into a pot. Set aside the flesh for later use. Add 3 cups of water to the syrup. Bring to the boil over high heat. Then turn to low heat and simmer until the syrup thickens. Put the pineapple flesh back in. Stir well. Keep on simmering over low heat for 5 more minutes until the syrup turns bubbly and the pineapple turns glossy. The jam is basically done at this stage if you like your jam to be thinner and runnier. Or, you may keep on cooking it longer if you prefer a stiffer jam.

4. Transfer the jam mixture into sterilized bottles while still hot. Seal them well for storage.

[Tips]

1. Please refer to "Passionfruit pineapple jam" on p.240.

[材料]

パイナップル...1000g（果肉のみ）
干し梅（中国話梅）...50g（約10コ）
ライム汁...1 コ
砂糖...250g
水...3杯

[作り方]

1. パイナップルを小口切り、干し梅の種を取って刻み、ライム汁、砂糖を混ぜ合わせ、ラップをし、冷蔵庫に入れ、水を出させるため一日置く。

2. 第二日に果肉と出た水分を鍋に入れ、弱火で沸騰してから、泡を取る。盛って冷ましてから冷蔵庫に入れ一日置く。

3. 第三日にストレーナーで果肉と汁を分け、汁を鍋に入れ、水3杯を加え、強火で沸騰してから弱火に汁が濃くなるまで煮、果肉を入れ混ぜ、汁を鍋に入れ、強火で沸騰してから弱火に汁が濃くなるまで煮、弱火で5分間に続けて煮る。泡が立ち始めるとき、果肉もピカピカになり始め、ジャム状になる。この時は基本的に出来上がる。自分の好みにジャムの濃度になるまで火を止める。短時間煮ると、汁が多くなり、長時間煮ると汁が濃くなる。

4. ジャムを盛り、熱いうち、消毒した容器に入れ、蓋をしてゆっくり召し上がる。

[ポイント]

1. 240ペッジ「パッションフルーツとパイナップルジャム」に参考する

The forgotten taste ~ My Auntie's pickled Ren Ren
忘れられた味わい － おばの仁稔（ヤンニム）..........P111

[Ingredients]

3 kg Ren Ren (a.k.a. Indochina Dragon Plum, or Ren Mian Zi)
white vinegar (enough to cover all Ren Ren)

[Marinade]

1.2 kg premium light soy sauce
1.2 kg sugar

[Method]

1. Rinse the Ren Ren and drain well. Make a crisscross incision on the skin with a knife. Then crush them gently with the flat side of a knife, so that they pick up the seasoning more easily.
2. Put the crushed Ren Ren into a tray. Add white vinegar and leave them for at least 4 hours. Make sure there is enough vinegar to cover all Ren Ren. Then rinse them with cold drinking water. Put them into a strainer to drain the water.
3. In a pot, put in light soy sauce and sugar. Cook over low heat until sugar dissolves. Leave this mixture to cool.
4. Put the Ren Ren into a container. Pour in the cooled soy sauce mixture. There should be enough soy sauce mixture to cover all Ren Ren. Leave them for 1 week and they are ready to be served. If you prefer some piquancy in your food, you may add some bird's eye chillies to the soy sauce mixture.

[材料]

仁稔（ヤンニム）...3000g
白酢...仁稔を浸せる量

[下味用つけ汁]

高級しょうゆ...1200g
砂糖...1200g

[作り方]

1. 仁稔を洗い、水気を取り、ナイフで表皮に十字の切り目を入れ、包丁の腹でたたいて味を染み込みやすくする。
2. 仁稔を漬け容器に入れ、白酢を入れ半日放置（少なくとも4時間）（白酢の量は必ず仁稔を浸せる）、その後は冷水で浸してからざるにあげて水分を切っておく。
3. フライパンにしょうゆと砂糖を入れ、弱火で溶けるまで煮、冷ましておく。
4. 仁稔を適当な容器に入れ、煮たしょうゆを加え（仁稔を浸せる）、漬けて一週間後に召し上がる。もし辛いのが好きなら、唐辛子を入れ一緒に漬け、風味がさらにアップ。

Tofu and thousand-year egg cold appetizer dressed in Ren Ren juice
仁稔（ヤンニム）汁とピータン豆腐の和え物..........P114

[Ingredients]

1 pack soft tofu for steaming
1 thousand-year egg
finely shredded spring onion

[Dressing]

1/2 cup Ren Ren juice
1 tbsp sesame oil

[Method]

1. Slice the tofu and arrange neatly in a deep dish. Finely dice the thousand year egg and arrange over the tofu. Drizzle with the dressing. Sprinkle finely shredded spring onion on top. Serve.

[Tips]

1. Sieve the Ren Ren juice before use so that the dressing looks clearer and nicer.

[材料]

絹ごし豆腐...1ボックス
ピータン...1コ
ねぎ（細かいせん切り）...適量

[涼拌ソース(リャンバンソース)]

仁稔汁（ヤンニム汁）...約1/2杯
ゴマ油...1大さじ

[作り方]

1. 豆腐をスライス切り、深い皿に盛り付け、ピータンを小さめに切って豆腐の上にのせ、涼拌ソースをかけ、ねぎを適量加える。

[ポイント]

1. 仁稔汁を使う前にストレーナーでかすを漉して、作った汁の見た目がもっときれいになる。

Traditional sweet and sour green papaya pickles
伝統の青パパイヤ甘酢漬け..........P120

[Ingredients]

1 green papaya (about 1 kg)
1 tsp salt

[Marinade]

500 ml white vinegar
400 g sugar

[Method]

1. Add sugar to vinegar. Cook over low heat until the sugar dissolves. Turn off the heat. Leave it to cool.
2. Peel the green papaya. Cut it into halves. Remove the seeds and slice it. Transfer into a bowl. Sprinkle salt on top and mix well. Leave it for half a day to draw the moisture out.
3. Drain any liquid in the bowl. Squeeze the papaya sliced dry. Transfer into a container. Pour in the cooled pickling vinegar from step 1. Mix well. Leave it for 1 to 2 days. Serve.

[材料]

青パパイヤ...1コ(約1000g)
塩...1小さじ

[下味付け汁]

白酢...500ml
砂糖...400g

[作り方]

1. 白酢に砂糖を入れ、弱火で砂糖が溶けるまで煮、冷ましておく。
2. 青パパイヤの皮を剥き、縦半分に切って種を取ってから、スライスして容器に入れて塩をかけ混ぜ、水を出させるため半日置く。
3. 青パパイヤから出た水分を捨て、水気を搾って容器に入れ、冷ました甘酢を入れ混ぜ、1〜2日に漬けてから出来上がる。

Passionfruit green papaya pickles
パッションフルーツとパパイヤの漬け物..........P122

[Method]

1. The method is the same as above, except that the pulp of 3 to 4 passionfruits is added to the pickling vinegar in step 3. Mix the passionfruit pulp well with the green papaya and pickling vinegar. The passionfruit would add a unique fruity fragrance to the pickles, making them even more appetizing.

[Tips]

1. Use green papayas of the right ripeness – it should yield slightly when pressed and its flesh should be pale yellow or pale orange in colour.

[作り方]

1. 作り方は上記と同じで、但し、作り方の手順3. に3〜4コのパッションフルーツ果肉を加え、甘酢と青パパイヤを混ぜ合わせる。パッションフルーツを加えたら、青パパイヤにパッションフルーツの特有の濃厚な香りを添え、食欲をそそる。

[ポイント]

1. パパイヤの丁度いい熟度を選んでください。パパイヤの表面を軽く押してあまり硬くならない程度、切り分けた後の果肉は浅い黄色または薄いオレンジ色が一番よい。

[Ingredients (A)]

2 cucumbers (about 1 kg, cut into strips)
1/2 white cabbage (about 500 g, sliced)
10-plus string beans (about 150 g, cut into short lengths)
2 carrots (about 500 g, sliced)
1 pineapple (about 1.5 kg, peeled and diced)
1/2 yam bean (about 500 g, sliced)
5 medium-sized green chillies (about 75 g, cut into rings)
5 medium-sized red chillies (about 75 g, cut into rings)

[Ingredients (B)]

toasted peanuts (chopped)
toasted sesames

[Marinade (A)]

50 g salt
1.5 kg sugar
1.5 litres white vinegar

[Marinade (B)]

600 g shallot (crushed)
200 g garlic (crushed)
8 candlenuts (crushed)
3 pieces turmeric (about 40 g, crushed)
2 cups oil
150 g chilli sauce for Hainan chicken rice (or garlic chilli sauce)

[Method]

1. Cut all ingredients (A) accordingly and place them into a container. Sprinkle salt on them. Mix well. Leave them overnight. Squeeze them dry and set aside.
2. Cook the sugar and vinegar in marinade (A) until sugar dissolves. Turn off the heat and set aside to let cool. Heat some oil in a wok. Stir fry the crushed ingredients in marinade (B) until dry and lightly browned. Add chilli sauce. Stir well. Set aside to let cool.
3. Mix marinade (A) with marinade (B). Put in the ingredients (A) and mix well. Leave them to soak for 1 day. Sprinkle toasted chopped peanuts and sesames on top. Serve.

[Tips]

1. To crush the aromatics in marinade (B), I prefer using a mortar and pestle. Try not to use a blender. A blender can only chop the ingredients up, while a mortar and pestle can bruise the ingredients and let their juice out completely. The pickles will then pick up their flavours better and taste better too.

[材料 (A)]

キュウリ...2本（約1000g）（千六本切り）
キャベツ...半分（約500g）（角切り）
いんげん...十本（約150g）（小さく段きり）
人参...2本（約500g）（スライス切り）
パイナップル...1コ（約1500g）（粗みじんきり）
葛芋...半分（約500g）（細切り）
ピーマン　中サイズ（緑・赤）...各5コ（約150g）
（輪きり）

[材料 (B)]

香ばしく炒めたピーナッツ(砕けたもの)...適量
香ばしく炒めた白ゴマ...適量

[下味用つけ汁 (A)]

塩...50g
砂糖...1500g
白酢...1.5リットル

[下味用つけ汁 (B)]

エシャロット...600g
にんにく...200g
キャンドルナッツ...約８コ
ウコン（ターメリック）...3本（約40g）
＊＊全部押し潰す
油...2杯
海南チキンライス用チリソース或いはにんにく入り
チリソース...約150g

[作り方]

1. 材料（A）を切ってから容器に入れ、塩を入れて
   よく混ぜ、一晩に置き、第二日に出た水を搾っ
   ておく。
2. 下味用つけ汁（A）に砂糖と酢をちょっとだけ煮
   て、砂糖が溶けたら火を止め、冷ましておく。
   フライパンに油を熱し、下味用つけ汁（B）エシ
   ャロット、にんにく、キャンドルナッツとみじ
   ん切ったウコンを香ばしく焼き、材料が乾いて
   きて黄色になるまで炒める。にんにく入りチリ
   ソースを入れ、混ぜ炒めてから取り出し、冷ま
   しておく。
3. 全部の下味用つけ汁をよく混ぜ合わせ、材料
   （A）を入れ、一日漬けてから完成する。召し
   上がる時に適量のピーナッツとゴマを振りかけ
   る。

[ポイント]

1. 下味用つけ汁（B）は最も石のすり鉢で潰す。な
   るべくミキサーを使用しないで、ミキサーは材
   料を潰すだけで、石のすり鉢は材料を細かく潰
   してエキスが完全に滲み出し、食材に完全に吸
   わせ、もっと美味しく出来上がる。

Traditional sweet vinegar pickled young ginger

[Ingredients]

2 kg young ginger
cold drinking water (enough to cover the ginger)

[Marinade]

1 litre white vinegar
800 g sugar
3 tsps salt

[Method]

1. Add sugar to white vinegar. Cook over low heat until sugar dissolves. Turn off the heat and leave it to cool.
2. Rinse the young ginger in cold drinking water. Drain. Slice and place into a container. Sprinkle salt on top. Mix well. Leave it for 1 day to drain the moisture out of the ginger.
3. Drain the ginger the next day. Soak it in cold drinking water for a while. Drain and squeeze dry the ginger.
4. Place the ginger into a container. Pour in the vinegar syrup from step 1. Mix well. Leave the ginger in the syrup for 3 days. Serve.

Soy marinated young ginger

[Ingredients]

2 kg young ginger
cold drinking water (enough to cover the ginger)

[Marinade (A)]

500 ml Kikkoman soy sauce
200 ml Guangdong rice wine
200 ml mirin (Japanese cooking wine)
350 ml cold drinking water
500 g sugar

[Marinade (B)]

3 tsps salt

[Method]

1. Mix all marinade (A) ingredients together. Cook over low heat until sugar dissolves. Turn off the heat. Leave it to cool.
2. Rinse the young ginger in cold drinking water. Drain. Slice and place into a container. Sprinkle salt (i.e. Marinade (B)) on top. Mix well. Leave it for 1 day to drain the moisture out of the ginger.
3. Drain the ginger the next day. Soak it in cold drinking water for a while. Drain and squeeze dry the ginger.
4. Place the ginger into a container. Pour in the marinade (A) from step 1. Mix well. Leave the ginger in the marinade for 3 days. Serve.

Plum-vinegar marinated young ginger sprouts

[Ingredients]

1 kg young ginger sprouts
(i.e. the youngest sprouts on a young ginger stem.)
cold drinking water (enough to cover ginger sprouts)

[Marinade]

500 ml plum vinegar
(Please refer to the method on p.221. Or, you may get plum vinegar in a bottle from Japanese supermarkets.)
sugar
3 tsps salt

[Method]

1. Add sugar to plum vinegar. Adjust the amount of sugar according to your preferred sweetness. Cook over low heat until sugar dissolves. Turn off the heat and leave it to cool. If you're getting ready-made plum vinegar in a bottle, taste it first to see if you need to add more sugar.
2. Rinse the young ginger sprouts in cold drinking water. Drain. Slice and place into a container. Sprinkle salt on top. Mix well. Leave it for 1 day to drain the moisture out of the ginger.
3. Drain the ginger sprouts the next day. Soak it in cold drinking water for a while. Drain and squeeze dry the ginger sprouts.
4. Place the ginger sprouts into a container. Pour in the plum vinegar syrup from step 1. Mix well. Leave the ginger in the syrup for 3 days. Serve.

[Tips]

1. Young ginger has a mild heat and is not too spicy. It has a crunchy texture without being fibrous. That's why it is a great ingredient to be pickled. Pickled young ginger can be paired with different food to accentuate its taste. For instance, a classic combination will be pickled young ginger with thousand-year eggs. Together with Ren Ren (a.k.a. Indochina dragon plum or Ren Mian Zi) and fermented soybean paste, pickled young ginger also makes a great condiment to be steamed with seafood or meat.
3. Pick young ginger that has pale and white skin, without dirt or mud clung on it. It should be plump in shape and smells mildly fragrant. Quality young ginger should not have any rotten part on it. Its sprouts should be in pinkish or purplish colour.

伝統的な新しょうがの甘酢漬け

[材料]

新しょうが...1200g
冷水,,,適量（新しょうが全身を浸せる量）

[下味用つけ汁]

白酢...600g
砂糖...800g
塩...3小さじ

[作り方]

1. 白酢に砂糖を入れ、弱火で砂糖を溶かしてから火を止め、冷ましておく。
2. 冷水で新しょうがを洗い、水気を取り、薄く切り、容器に入れ、塩を入れ混ぜ、水を出させるため一日置く。
3. 翌日に新しょうがから出た水分を取り除き、冷水にちょっと浸し、水を切り、新しょうがを手で絞り水気を切る。
4. 最後に水気を絞った新しょうがを容器に入れ、先ほど砂糖を溶かした酢に入れよく混ぜる。漬けて約3日後には召し上がれる。

新しょうがの醤油漬け

[材料]

新しょうが...1200g
冷水...適量（新しょうが全体が浸る量）

[下味用つけ汁（A）]

キッコマン醤油...500ml
広東米酒（カントンミーチュウ）...200ml
みりん...200ml
冷水...350ml
砂糖...500g

[下味用つけ汁（B）]

塩...3小さじ

[作り方]

1. 下味用つけ汁（A）を混ぜ合わせ、弱火で砂糖を溶かしてから火を止め、冷ましておく。
2. 手順2,3は伝統的な新しょうがの甘酢漬けと同じ。
3. 最後に水気を絞った新しょうがを容器に入れ、冷ました下味用つけ汁（A）を入れ混ぜ、漬けて約3日後には召し上がれる。

葉しょうがの梅酢漬け

[材料]

葉しょうが...1200g（葉しょうがは新しょうがから出た芽の部分。）
冷水...適量（新しょうがの全身を浸る量）

[下味用つけ汁]

梅酢...500ml
（梅酢の作り方は221ページの「お酢でヘルシー ー 自家製梅酢」、または日本の大型スーパーで買える）
砂糖...適量
塩...3小さじ

[作り方]

1. 梅酢に砂糖を入れ、自分の好みに甘さを調節し、弱火で砂糖を溶かしてから火を止め、冷ましておく。直売所で買った梅酢の場合は梅酢の甘さを試し、砂糖を加える必要かどうか又は加える砂糖の量を決定する。
2. 伝統的な新しょうの甘酢漬けの手順2,3と同じ
3. 最後に水気を絞った葉しょうがを容器に入れ、先ほど出来上がった梅酢を入れ混ぜ、漬けて約3日後には召し上がれる。

[ポイント]

1. 新しょうがはさわやかな辛味、繊維が柔らかく、サクッとした歯ごたえ、漬けるのがよく合う。漬け上がった新しょうがはいろいろな食材に合せると、もっと美味くなる。皮蛋（ピータン）と新しょうがは外せない仲間だろう。それに、仁稔(ヤンニム)、と味噌を加え、いろいろな海鮮や肉などを蒸すのが食欲がそそられてくる。
2. 新しょうがの選び方は根の分部が白く、ふっくらしてみずみずしい、表面がしっとりとしていて、しょうがの香があり、傷がなく、茎の付け根が鮮やかな紅色のものを選ぶようにしましょう。

Winter melon fermented soybean paste
冬瓜味噌..........P129

[Ingredients]

2 kg winter melon
3 tsps salt

[Marinade]

350 g Liu Ma Kee fermented soybean paste
60 g sugar
1/2 cup rice wine
10 g liquorice

[Method]

1. Peel and seed the winter melon. Cut into cubes about 1.5 cm on each side. Put them into a large tray. Add salt and mix well. Leave it for 1 day so that water is drawn out of the melon.
2. Drain any liquid in the tray. Squeeze the diced winter melon dry.
3. Mix the marinade well. Put in the squeezed diced winter melon. Mix well and seal with cling film. Leave it in the fridge for 3 days. Stir the winter melon once or twice throughout the marinating process to ensure even infusion.

[材料]

冬瓜...2リットル
塩...3小さじ

[下味用つけ汁]

中国の味噌（廖孖記）...350g
砂糖...60g
米酒...1/2コップ
甘草...10g

[作り方]

1. 冬瓜の皮を剥き、種を取り、1.5cm角のさいの目切り、大きい容器に入れ、塩を加え、水を出させるため一日置く。
2. 第二日に冬瓜から出た水を取り除き、手で冬瓜の水分を搾り出す。
3. 下味用つけ汁をよく混ぜて、冬瓜を入れ混ぜ、ラップをし、冷蔵庫に入れ、約三日間に漬けてから出来上がる。漬ける間に時々一回か二回ぐらい混ぜ合わせ、味をよく染み込ませる。

Steamed pomfret with
winter melon fermented soybean paste
冬瓜味噌入りマナガツオ蒸し........P132

[Ingredients]

1 pomfret (about 600 g)
4 to 5 tbsps winter melon fermented soybean paste
shredded spring onion
2 to 3 tbsps oil

[Method]

1. Dress and rinse the pomfret. Make a crisscross incision on the fleshiest part of each side of the fish. This way, the fish get steamed more evenly and it picks up the seasoning better.
2. Put the fish on a steaming plate. Spread the winter melon fermented soybean paste on the fish. Boil water in a wok or steamer. Steam over high heat for 10 – 12 minutes. Sprinkle shredded spring onion on top. Dribble smoking hot oil over the fish. Serve.

[材料]

マナガツオ...1本(約600g)
冬瓜味噌...約4〜5大さじ
ねぎ（せん切り）...適量
油...約2〜3大さじ

[作り方]

1. マナがツオをおろし、よく洗い、全体を均一に蒸し上がるには味が染み込むように、魚の両面に十字に切り込みを入れる。
2. 冬瓜味噌はマナガツオの上にかけ、水が沸騰してから約10〜12分間に蒸し、取り出し、ねぎをかけ、熱した油をかけて出来上がる。

[Ingredients]

300 g ground pork
3 tbsps winter melon fermented soybean paste

[Marinade]

1 tbsp light soy sauce
1/2 tsp sugar
1/2 tsp chicken bouillon powder
1 egg white
1/2 tsp caltrop starch

[Method]

1. Add the marinade to the ground pork. Mix well. Stir with a chopstick in one direction until sticky. Then roll it into a patty. Slap the pork patty on a chopping board repeatedly until resilient. Place the pork patty on a steaming plate. Leave it for 1 hour.

2. Spread winter melon fermented soybean paste on the pork patty. Boil water in a wok or steamer. Steam over high heat for 12 to 15 minutes until done (depending on the thickness of the patty). You may sprinkle finely shredded spring onion on top if you want.

[Tips]

1. Steamed pork patty is a very basic home-style dish. That being said, there are a few tricks to make it fluffy and tender. First of all use pork shoulder butt meat for pork patty. It's because it is a very soft cut of meat that guarantees fluffiness after being steamed. You should also chop the pork up with a knife instead of using a grinder or food processor. The second trick works but maybe a bit unhealthful – add some fatty pork into the lean pork. The fat actually keeps the pork moist and the steamed patty will taste smoother. In fact, the tender and smooth pork patties you get from restaurants all have fatty pork stirred in.

[材料]

豚挽き肉...300g
冬瓜味噌...約3大さじ

[下味用つけ汁]

醤油...1大さじ
砂糖...1/2小さじ
チキンパウダー...1/2小さじ
卵白...1コ
片栗粉...1/2小さじ

[作り方]

1. 豚挽き肉に下味用つけ汁を入れてよく混ぜ、お箸で一方向によく混ぜ、そして、粘りがでるまで手づかみしてボウルに投げ付けて、蒸し皿に肉を盛り、約一時間下味をつける。

2. またひき肉の上に冬瓜味噌を塗り付け、水が沸騰してから、強火で12～15分ほど蒸し（蒸し時間は肉の厚さにより）自分の好みにみじん切りにしたねぎを上にかけて召し上がる。

[ポイント]

1. ひき肉蒸しは普通の家庭料理で、滑らかに口の中で肉がひろがる感じて、作り技がある。①肩ロースを選び、肉質が柔かいので蒸し上がったひき肉が柔かくなる。②このアドバイスは健康に対してよくなさそう。それは脂肪を加え、こうしたら、蒸した肉が滑らかになり、実は外で食べられるひき肉蒸しは同じ方法で脂肪を加えてある。

Thai pickled green mango
タイ風の青マンゴー漬け..........P136

[Ingredients]

3 Thai green mangoes (about 400 g)
200 g sugar
1 tsp salt

[Method]

1. Rinse the green mangoes and peel them.
2. Slice the green mangoes thickly. Place them into a bowl. Add sugar and salt. Mix well. Cover with cling film. Keep in the fridge for 1 day.
3. Water will be drained from the mangoes the next day and the mangoes turns soft. Stir the mangoes in the syrup well. Leave them in the fridge for 1 more day.

[Tips]

1. Green mangoes are available from Thai grocery stores. Pick those unripe ones that are hard to the touch. Only unripe mangoes will taste crunchy after being pickled.
2. Pickled green mangoes last in the fridge for months. It is a great summer snack that whets the appetite and refreshes your mind. You may also use it as a condiment to stir fry with beef. It is an innovative and tasty dish.
3. The syrup carries the juice drawn out of the mangoes. Add some lime juice and soda water for a refreshing summer drink.

[材料]

タイ産の青マンゴー...3コ（約400g）
砂糖...200g
塩...1小さじ

[作り方]

1. 青マンゴーを洗い、皮を剥く。
2. マンゴー果肉を厚切りにし、大きい容器にのせ、砂糖を入れ混ぜる。ラップをし、冷蔵庫に入れ一日置く。
3. 第二日にマンゴー果肉から水分が出るによって軟化する。マンゴー果肉と水をかき混ぜ、再び冷蔵庫に入れて一日置いてから召し上がる。

[ポイント]

1. 青マンゴーはタイのスーパーで買える。生っぽいと硬めを選び、漬けたマンゴーはシャキシャキした感じで美味しい。
2. 漬けたマンゴーを冷蔵庫に入れ、数ヶ月保存できる。食欲増進させる夏にぴったりの一品。牛肉と炒めて、美味しくて新しい発想の料理。
3. マンゴーから出た汁にライム汁、ソーダ水を加え、さっぱりとした夏のスペシャルドリンクになる。

Stir-fried beef with Thai pickled green mangoes
タイ風の青マンゴー漬けと牛肉炒め..........P138

[Ingredients]

200 g Thai pickled green mangoes
200 g beef tenderloin (sliced)
1/2 green bell pepper (sliced)
1/2 red bell pepper (sliced)

[Marinade]

1/2 tsp salt
1 tsp chicken bouillon powder
1/2 tsp sugar
1 tsp light soy sauce
1/2 tsp dark soy sauce
1/2 egg
50 g water

[Thickening glaze: (mixed well)]

1 tsp caltrop starch
water

[Method]

1. Add marinade to the beef. Mix well and leave
   it for 30 minutes. Set aside.
2. Heat a wok and add oil. Stir fry beef until
   medium done. Set aside. Heat the wok and
   add oil again. Stir fry green and red bell
   pepper briefly. Put the beef back in and stir
   well. Lastly put in the pickled green mangoes.
   Stir well. Stir in the thickening glaze at last. Mix
   well. Serve.

[材料]

青マンゴーの漬物...２００ｇ
牛ヒレ...２００ｇ(スライス切り)
ピーマン　中サイズ（緑・赤...各半個（スライス
切り）

[下味用つけ汁]

塩...1/2小さじ
鶏がら粉...１小さじ
砂糖...1/2小さじ
醤油...１小さじ
老抽（ラオ　チョウ）...1/2小さじ
タマゴ...半分
水...５０ｇ

片栗粉...１小さじ（とろみをつける用）
水...適量

[作り方]

1. 牛肉は下味用つけ汁を混ぜ合わせ、３０分を漬
   けておく。
2. フライパンに油を入れ、熱し、牛肉を入れ、６
   割ほど炒めたら取り出す。再びにフライパンを
   熱し、緑と赤のピーマンを入れ、軽く炒め、牛
   肉をフライパンに戻し入れ、青マンゴーを入れ
   て炒める。最後に水溶き片栗粉を入れ、とろみ
   をつけてから出来上がる。

[Ingredients]

1 Thai green mango
3 sprigs fresh coriander
1 clove shallot
1 bird's eye chilli (adjust the amount according to your preferred piquancy)
toasted peanuts
8 sand prawns (Gei Wai prawns)
1/2 small squid

[Salad dressing]

5 tbsps lime juice
2 1/2 tbsps fish sauce
3 1/2 tbsps sugar

[Method]

1. Rinse all ingredients. Peel the green mango and slice it. Cut the coriander into short lengths. Cut the shallot and chilli into rings. Grind the peanuts. Blanch the prawns in boiling water and shell them. Blanch the squid in boiling water and cut into rings. Mix the salad dressing well and set aside.
2. Put all ingredients (except the ground peanuts) into a salad bowl. Add salad dressing and toss well. Transfer onto a serving plate. Sprinkle ground peanuts on top and serve.

[Tips]

1. Green mangoes are available from Thai grocery stalls in Wanchai Market (at the corner of Queen's Road East and Stone Nullah Lane) or Kowloon City.
2. Sprinkle ground peanuts on the salad right before you serve. If you put them in too early, they'd pick up the moisture in the dressing and become soggy.

[材料]

タイの青マンゴー...1コ
新鮮なコリアンダー...3本
エシャロット...1粒
唐辛子...1つ(好みの辛さに加減する)
香ばしく炒めたピーナッツ...適量
エビ...約8匹
Sサイズのイカ...半匹

[ドレッシング]

ライム汁...5大さじ
ランプラー...2 1/2大さじ
砂糖...3 1/2大さじ

[作り方]

1. 全部の材料を洗い、青マンゴーの皮を剥き、スライスする。コリアンダーを段切り、エシャロットと唐辛子を輪切りにし、ピーナッツを押し潰し、エビはボイルして殻を剥き、イカをボイルして輪切りにし、ドレッシングを混ぜておく。
2. 全部の材料を容器に入れ(ピーナッツを除き)、ドレッシングを入れ混ぜ、皿に盛る。上に押し潰したピーナッツをふりかけて召し上がる。

[ポイント]

1. タイの青マンゴーは「湾仔道石水渠街市」、「九龍城」などのタイ系スーパーマーケットで売られる。
2. ピーナッツは必ず最後にふりかけ、早く入れるとピーナッツがサラダの水分を吸収し、カリっとした食感にならない。

White bitter melon marinated in Okinawa Kurozatou
白ゴーヤの沖縄黒砂糖漬け..........P142

Pickling vegetables in sugar actually helps remove the grassy taste of them. Those easily perishable vegetables with strong flavour or bitterness especially benefit from sugar pickling. The sugar also brings out the enzymes and micro organisms in the vegetables, so that the pickles taste sweeter and crunchier.

[Ingredients]

2 white bitter melons (about 850 g)
3 tsps salt

[Marinade]

150 g Okinawa Kurozatou (black sugar)
20 g Zhenjiang black vinegar
30 g ginger juice

[Method]

1. Rinse the bitter melons. Cut into halves along the length. Seed and slice them. Then remove the mesh-like tissue in the fruit. Put them into a container. Add salt. Mix well and leave them for 1 day.
2. Blanch the bitter melons in boiling water. Then plunge them immediately into iced water. Wait until they are cooled completely. Drain. Squeeze the bitter melons dry.
3. To make the marinade, crush the Kurozatou. Mix well with black vinegar and ginger juice. Put in the bitter melons. Mix well. Leave them for 2 to 3 days. Serve.

[Tips]

1. The bitterness of bitter melon mostly comes from the mesh-like tissue inside the fruit. If you cut it off, it would taste less bitter. But if you prefer the characteristic bitterness of bitter melon like myself, you might keep the mesh-like tissue. After all, it is the most nutritious part of a bitter melon.

砂糖で漬物を漬けると、野菜の青臭いを取り除く。味が濃く、腐りやすく、渋みや苦味のある野菜に合う。そして、野菜の水分中に酵素と微生物を引き出せ、漬物が甘みと爽やかな味になる。

[材料]

白ゴーヤ...2コ （約850g）
塩...3小さじ

[下味用つけ汁]

沖縄産黒砂糖...150g
黒酢...20g
じょうが汁...30g

[作り方]

1. ゴーヤを洗い、縦半分切り、種とワタを取り除き、横に薄切り、容器に入れ、塩をふってもみ、一日に置く。
2. 第二日にゴーヤを湯引きし、すぐに氷水に浸し、冷ましてから取り出し、水分を絞る。
3. 黒砂糖を潰し、黒酢とじょうが汁と混ぜ合せ、ゴーヤを入れ混ぜ、約2～3日漬けた後に召し上がる。

[ポイント]

1. ゴーヤの苦い部分は主にワタ。その部分を取り除ける。もし僕と同じゴーヤの苦味が好きであれば、そのままワタを残してよい。更に、この部分がゴーヤの一番栄養がある。

Homemade apple vinegar
自家製リンゴ酢..........P146

Japanese Ourin apple vinegar
*small batch

[Ingredients]

2.3 kg Japanese Ourin apples (about 10 apples)

[Marinade]

2.5 litres white vinegar
500 g maltose
800 g rock sugar

Japanese Fuji apple vinegar
*large batch

[Ingredients]

4.6 kg Japanese Fuji apples (about 20 apples)

[Marinade]

5 litres white vinegar
1 kg maltose
1.6 kg rock sugar

[Method]

1. Rinse the apples. Wipe dry with paper towel completely. Leave them at room temperature for a few hours to make sure all surface moisture has evaporated. Prick some holes on the apples with a bamboo skewers. That helps the apple juice to be infused into the vinegar.
2. Sterilize a container. Put in the apples. Pour in white vinegar, maltose and rock sugar. Seal the lid well. Leave them at a cool dry place away from the sun. Shake the container every once in a while to enhance infusion. It takes at least 6 months for the vinegar to be flavourful. When the apples have shrunken and wrinkled up, the vinegar is ready to be served.

[Tips]

1. The surface of the apples MUST be completely dry for the vinegar to have a long shelf life. Otherwise, the vinegar turns stale easily.

以下の2種類リンゴ酢は製作分量、リンゴの品種が異なりますが、同じく濃厚芳醇な味わいを持ち、美味しくて魅力がある。

王林リンゴ酢
（量は少ない）

[材料]

日本の王林リンゴ...2300g（約10コ）

[下味用漬け汁]

白酢...2.5リットル
モルトシュガー...500g
氷砂糖...800g

富士リンゴ酢
（量は多い）

[材料]

日本富士リンゴ...4.6リットル（約20コ）

[下味用漬け汁]

白酢...5リットル
モルトシュガー...1000g
氷砂糖...1600g

[作り方]

1. リンゴを洗い、キチンペーパーで水気を拭き取り、残りの水分を完全に切るように数時間に放置する。汁を滲み出せるため竹串でリンゴに小穴を刺す。
2. 消毒した容器にリンゴを入れ、白酢、モルトシュガーと氷砂糖を加え、蓋をし、冷暗乾燥と直射日光が当たらない場所に保存し、時々容器を動き揺らして、少なくとも6ヶ月に漬け、リンゴの皮は皺が寄り、体積が減ると出来上がる。

[ポイント]

1. 必ずリンゴの表面に水気を完全に切る、こうすると、リンゴ酢が長時間保存できる。もし、漬けるときに、生水が混ざっているとリンゴ酢が腐りやすくなる。

Pickled cherry tomatoes in apple vinegar
ミニトマトのリンゴ酢漬け..........P149

[Ingredients]

1 litre homemade apple vinegar
1 kg cherry tomatoes

[Method]

1. Rinse the cherry tomatoes. Make a crisscross incision lightly on the bottom of each cherry tomato. Boil some water in a pot. Blanch the cherry tomatoes in boiling water for 10 to 20 seconds until the skin starts pulling away slightly along the cuts. Drain and soak the cherry tomatoes immediately in iced water. Then peel the skin off carefully with a small knife. Drain them well and set aside.
2. Place the cherry tomatoes into a container. Pour in enough apple vinegar to cover the cherry tomatoes. Seal the container and leave them for 1 day. Serve.

[材料]

リンゴ酢...1リットル
ミニトマト...約1kg

[作り方]

1. ミニトマトを洗い、まずはナイフでミニトマトの底に十字の切り込みを入れ、お湯を沸かしてからミニトマトを入れ、十数秒ほどさらっと茹で、皮がちょっと破れてきたら取り出してすぐに氷水に取り、ナイフでミニトマトの皮を剥き、水気を切っておく。
2. 容器にミニトマトを入れ、リンゴ酢を入れ(リンゴ酢が必ずトマトの表面を覆う)、蓋を閉め、約一日につけてから召し上がる。

Cherry tomato vinegar jelly
ミニトマト酢のゼリー..........P152

[Ingredients]

1 litre apple vinegar that has been used to pickle cherry tomatoes
50 g gelatine leaves (about 10 leaves)

[Method]

1. Soak the gelatine leaves in iced water for about 10 minutes until soft. Squeeze dry. Set aside.
2. Heat the apple vinegar up to 60 to 70℃. Put in the gelatine leaves. Heat the mixture while stirring constantly until the gelatine dissolves completely. Strain the resulting mixture to remove impurities and any gelatine residue.
3. Pour the strained gelatine mixture into individual jelly cups or a flat tray. Leave it to cool. Refrigerate until set.

[材料]

ミニトマトの酢...1リットル
板ゼラチン...50g(約10枚)

[作り方]

1. 板ゼラチンは氷水に10分ほどつけ柔らかくなったら、水気を絞っておく。
2. ミニトマトの酢は60℃〜70℃ぐらい加熱し、板ゼラチンを入れ、完全に溶けるまで煮ながら攪拌し、目の細かい漉し器で漉して雑質や板ゼラチンの溶かし残しを取り除く。
3. 容器にミニトマトの酢を入れ、冷ましてから冷蔵庫に入れて固くなるまで冷やし、ゼリー状になって出来上がる。

Cherry tomato vinegar drink
ミートマト酢のドリンク..........P154

Chill the apple vinegar that has been used to pickle the cherry tomatoes. Transfer vinegar into a small glass. Add a pickled cherry tomato. Serve.

漬け上がったミニトマトの酢を冷やしてからコップに入れ、ミニトマト1粒を加えて召し上がる。

Eggplant pickled with maggi's seasoning and balsamic vinegar
ナスのマギーと黒酢漬け..........P184

[Ingredients]
2 Japanese eggplants

[Marinade]
150 g Maggi's seasoning
50 g Balsamic vinegar
60 g sugar
60 ml sesame oil

[Method]
1. Rinse the eggplants. Boil water in a wok or steamer. Steam them for 15 minutes until soft and done. Leave them to cool. Cut each eggplant into halves along the length.
2. Mix the marinade well. Put the eggplants into the marinade and leave them overnight. Flip the eggplant upside-down once in a while throughout the pickling process to ensure they pick up the flavours evenly. Transfer eggplants on serving plate. Drizzle with marinade. Serve.

[材料]
日本産のナス...2本

[下味用漬け汁]
美極鮮醤油（マギーソース）...150g
イタリア産の黒酢...50g
砂糖...60g
ゴマ油...60ml

[作り方]
1.ナスを洗い、水が沸騰してから15分間にナスを煮、取り出して盛り、冷めてから縦半分に切っておく。
2.下味用漬け汁を混ぜ、ナスを入れ、一晩につけて召し上がり、漬ける間にナスを丁寧に動かして、味を均一にしみこませる。召し上がる時にナスを皿に盛り、適量な下味付け汁をかけて召し上がる。

Low-temperature air-dried tomatoes with truffle oil
ドライトマトのトリュフオイル入り..........P155

[Ingredients]

10-plus medium tomatoes
(roma or golden banana variety)

[Seasoning]

truffle oil
salt
ground white pepper

[Method]

1. Rinse the tomatoes and wipe dry. Cut into halves. Preheat an oven to 60℃. Line a baking tray with aluminium foil. Put on the tomatoes. Sprinkle some salt and ground white pepper over them. (Optionally, sprinkle some sugar on them if they are too tart.) Bake in the oven for 6 hours until the tomatoes are half-dried. Drizzle with truffle oil. Bake for 30 more minutes.

2. Leave the tomatoes to cool completely. Store in airtight boxes.

[Tips]

1. When you bake the tomatoes in an oven, so that the steam can escape. Otherwise, the steam will be trapped in the oven and the tomatoes will not be dry enough.

2. Truffle oil should be added at last. It's because oil tends to seal in the moisture, so that it takes longer for the tomatoes to get dry. Furthermore, the unique delicate fragrance of truffle oil is easily destroyed after prolonged heating.

[材料]

トマト（中）...約十数コ
（ローマトマト或はプラムトマト）

[調味料]

トリュフオイル、塩、コショウ...各適量

[作り方]

1. トマトを洗い、水気を取り、縦半分に切り、オーブンで60度予熱し、オーブン皿にホイルを敷き、トマトをのせ、塩とコショウを少々振りかけ、（もし買って来たトマトが酸っぱいであれば、砂糖を少々加えてもよい）オーブンで6時間ほど乾燥させ、トマトが半乾燥の状態に適量なトリュフオイルをかけ、後半時間で乾燥させてから出来上がる。

2. オーブンからトマトを取り出し、冷めてから、容器に保存してゆっくり召し上がる。

[ポイント]

1. 乾燥する時にオーブンのドアを少し開け、蒸気を出させ、オーブンの中に蒸気がこもるとトマトが十分に乾燥させられない。

2. トリュフオイルは必ず最後に入れ、そうしなければトマトの水分が蒸発できなくなり、もっと長時間に乾燥し、トリュフオイルが長時間に加熱すると香味が飛んでしまう。

Cold tomato appetizer in plum dressing with mint leaves

冷やしミント、梅、トマト..........P158

[Ingredients]

5 to 6 tomatoes
fresh mint leaves

[Dressing (A)]

40 g dried plums (about 10-plus plums)
40 g rock sugar (crushed)
1 tbsp lime juice (juice of one lime)
1 cup water

[Dressing (B)]

1 tbsp balsamic vinegar
olive oil
2 tsps plum powder

[Method]

1. Boil the dressing (A). Leave it to cool. Put in the balsamic vinegar and olive oil from dressing (B). Mix well. Keep in the fridge for later use.
2. Blanch the tomatoes in boiling water for about 10 - 15 seconds. Drain and dunk them into iced water. The thermal shock will break the skin. Peel the tomatoes. Refrigerate until cold.
3. Before serving, slice the tomatoes across their height. Sandwich a mint leaves between the slices. Dribble the chilled dressing from step 1 over the tomatoes. Sprinkle plum powder on top. Serve.

[Tips]

1. Alternatively, you may put the sliced tomatoes and mint leaves into the dressing. Leave them in the fridge for 2 to 3 hours until the tomatoes pick up the dressing flavour. Sprinkle plum powder and serve. This is an awesome summer appetizer.

[材料]

トマト...5〜6コ
新鮮なミントリーフ...適量

[下味用漬け汁 (A)]

干し梅（話梅）...40g(約十数個)
氷砂糖(砕けたもの)...40g
ライム汁...1大さじ（約1コ分）
水...1杯

[下味用漬け汁 (B)]

イタリア産の黒酢...1大さじ
オリーブオイル...適量
梅粉...約2小さじ

[作り方]

1. 下味用漬け汁（A）を煮、冷めてから下味用漬け汁（B）の黒酢とオリーブオイルを入れ混ぜ、冷蔵庫に入れておく。
2. トマトを沸騰したお湯に入れて約十数秒後に、すぐ取り出して氷水に浸し、表皮が裂けると、皮を剥き、冷蔵庫に入れて冷やしておく。
3. 召し上がる時にトマトを横にスライス切り、スライスしたトマトの間にミントリーフを挟み、下味用漬けソースにかけ、適量な梅粉を振りかけて召し上がる。

[ポイント]

1. 別の方法で召し上がるのは、トマトスライスとミントリーフに調節した下味用漬け汁を入れ、冷蔵庫に2〜3時間に下味つけてから皿に盛り、梅粉を振りかけて召し上がる。たまらない夏の逸品だろう。

Homemade foolproof five-spice salted eggs
自家製香辛料と塩漬けの卵～「失敗ゼロ」簡単な自家製料理..........P160

[Ingredients]

10 duck eggs
10 chicken eggs
2 litres water
500 g coarse salt
10 g Sichuan peppercorns
10 g star anise
10 g Tsaoko fruit
10 g cloves
10 g cassia bark

[Method]

1. Rinse the eggs and wipe dry.
2. In a pot, boil the water with salt and spices until the water is infused with flavour. Stir well until the salt dissolves. Leave this soaking brine to cool completely.
3. Place the eggs into the cooled soaking brine. Seal with cling film. Leave them to soak for 1 month. The eggs can then be served.

[Tips]

1. As the soaking brine is highly concentrated, the eggs tend to float in it. The part of an egg that sticks out of the brine might not pick up the salt and flavours properly. In this case, you may fill your bowl fully with the brine. Put in the eggs and cover the bowl with a chopping board (given the chopping board is large than the mouth of your bowl) topped with some heavy weights. In this case, all eggs will be pressed beneath the brine surface and salted evenly. If this method doesn't work in your case, you'd have to stir the eggs once a week to ensure they are salted evenly.

[材料]

アヒル卵・鶏卵...各10コ
水...2リットル
粗塩...500g
花椒、八角、草果、丁香、桂皮...各10g

[作り方]

1. アヒル卵、鶏卵を洗い、水気を取る。
2. 水、塩、香料を煮て味が出させる。粗塩が完全に溶けるまで攪拌し、冷ましておく。
3. 水分が完全に冷めてから、アヒル卵、鶏卵を入れ、ラップをし、静止した状態で約1ヶ月放置してから出来上がる。

[ポイント]

1. 水の中に塩分濃度が高いので、浮力が強く、タマゴを入れてから、一番上層のタマゴは浮かべていて味付けられない。こうした場合は、塩水をいっぱいまで注ぐ、まな板をのせ、上に重しをのせる。こうすれば、タマゴは完全に塩水に浸り、味が均一に浸み込み。もし、このやり方ができなければ、一週間おきにタマゴを軽く攪拌し、タマゴに味を吸収させる。

Stir-fried bitter melon with five-spice salted eggs
香辛料塩漬け卵とゴ　ヤ炒め..........P164

[Ingredients]
2 bitter melons (about 400 g)
2 homemade five-spice salted eggs
3 cloves garlic (gently crushed)
1/2 red chilli

[Seasoning]
1/3 tsp salt
1/2 tsp sugar
1/3 tsp chicken bouillon powder

[Method]
1. Rinse the bitter melons and seed them. Slice and blanch in boiling water. Drain. Set aside. Steam the salted eggs for 15 minutes until done. Shell and dice them. Crush the garlic cloves with the flat side of a knife. Slice the red chilli.
2. Heat oil in a wok. Stir fry garlic until fragrant. Add bitter melon and red chilli. Stir well. Put in diced salted eggs and seasoning. Stir well and serve.

[Tips]
1. Before blanching the sliced bitter melons, you may add 1/2 tsp of salt to them and mix well. Leave them for 30 minutes until moisture is drawn out of the bitter melons. Then squeeze them dry and blanch them in boiling water. The bitter melon will taste less bitter and crunchier this way.

[材料]
ゴーヤ...2本(約400g)
自家製五香シェンタン（塩漬け卵）...2コ
にんにく...約3つ（押し潰す）
赤パプリカ...半個

[調味料]
塩...1/3小さじ
砂糖...1/2小さじ
チキンパウダー...1/3小さじ

[作り方]
1. ゴーヤを洗い、種を取り除き、薄切り、湯切りして水気を切る。シェンタンを約15分で蒸し、粗みじん切り、にんにくを押し潰し、赤パプリカを角切っておく。
2. フライパンに油を熱し、にんにくを入れて香ばしく焼き、ゴーヤと赤パプリカを混ぜ炒め、シェンタンと調味料を入れて混ぜ合わせ、皿に盛って召し上がる。

[ポイント]
1. ゴーヤは約1/2小さじ塩を入れ混ぜ、水を出させるため約30分置く、水気を絞り、湯切りする。こうすると、苦みを抑え、炒めたゴーヤはもっとシャクシャクした感じ。

Salmon Shoyuzuke with sake and dill
サーモンの清酒、ハーブ、醤油漬け..........P166

[Ingredients]

1 piece fresh salmon fillet (about 500 g)
4 to 5 sprigs fresh dill

[Marinade]

250 ml sake (Japanese rice wine)
250 ml mirin (Japanese sweet cooking wine)
250 ml Kikkoman soy sauce
about 1 1/2 tbsps honey

[Method]

1. Put sake and mirin into a pot. Turn on the heat. When it starts to boil, carefully tilt the pot so that it catches fire. Let it burn until it extinguishes by itself. Remove from heat and leave it to cool. Mix in the soy sauce and honey. Refrigerate for later use.
2. Put the salmon into the marinade that has been completely cooled. Seal with cling film. Leave it to soak for 1 day.
3. Drain the salmon. Put it on a cling film. Finely chop the dill. Arrange on the salmon. Wrap it up and leave it in the fridge for one more day. Serve.

[Tips]

1. Make sure you chill the marinade completely before you put in the salmon. The salmon served this way must be kept at a low temperature. If the marinade is warmer than the salmon, bacteria will grow and the fish goes stale easily.

[材料]

新鮮なサーモン...1枚（約500g）
ディル...4〜5本

[下味用漬け汁]

日本清酒...250ml
みりん...250ml
キッコマン醤油...250ml
ハチミツ...約1 1/2大さじ

[作り方]

1. フライパンに清酒とみりんを入れ、沸騰し始める時に、酒の表面に火をつけ、火が消えたらコンロの火を止める。冷めてからキッコマン醤油とハチミツを入れ混ぜ、冷蔵庫に入れて冷やす。
2. 下味用漬け汁が完全に冷やしてから取り出し、サーモンを入れ、ラップをし、再び冷蔵こに入れて一日漬け。
3. 第二日にサーモンだけを取り出し、ソースをこして、ラップの上にのせ、ディルをみじん切り、サーモンにかけ、ラップをし、また一日漬けてから召し上がる。

[ポイント]

1. 下味用漬け汁は必ず冷やしてからサーモンを入れ、サーモンは基本的に冷凍保存し、漬け汁の温度がサーモンより高ければ、サーモンが細菌が繁殖し腐りやすくなる。

Pepper-scented salted salmon
サーモンのコショウ塩漬け..........P170

[Ingredients]

1 piece fresh salmon (about 500 g)

[Marinade]

2 tbsps sake (Japanese rice wine)
2 tbsps white peppercorns
1 tbsp premium sea salt

[Method]

1. Wipe dry the salmon with paper towel. Then brush sake evenly on all sides. Leave it in the fridge for 2 hours.
2. While the salmon is being marinated, crush the peppercorns in a mortar and pestle. Stir fry crushed peppercorns and sea salt in a dry wok until lightly browned and fragrant. Set aside to let cool. Rub the pepper-salt mixture on one side of the salmon. Wrap it in cling film. Keep in the fridge for 1 day. Cut into pieces and serve.

[Tips]

1. Just rub the pepper-salt mixture on one side of the salmon. If you coat the salmon in the mixture on all sides, it will be too salty. Besides, the salt will draw moisture out of the fish. If you put too much salt on it, the salmon will end up too wet. Thus, both the taste and the presentation will be adversely affected.

[材料]

新鮮なサーモン...1枚（約500g）

[下味用漬け汁]

日本清酒...2大さじ
粒コショウ...2大さじ
高級海塩...1大さじ

[作り方]

1. サーモンをキチンペーパーで水気を拭き取り、上に清酒を均一にかけ、冷蔵庫に入れ約2時間置く。
2. サーモンを漬ける間に、粒コショウを石のすり鉢で潰して、海塩と弱火で黄色になり始め香りが出るまで炒める。冷めてからサーモンの上に塗り付け、ラップをし、冷蔵庫に入れ一日置く、第二日ラップを取り外し、適当な大きさに切り分けて召し上がる。

[ポイント]

1. コショウと塩はサーモンの片側に塗り付けは十分です。もし、全体に塗り付けたら、しょっぱい過ぎになる。さらに、塩分が食材に合せると水が出る。完成品が水分が多過ぎで、味わいに影響するし見た目が良くない。

Pepper-scented salted chicken
鶏のコショウ入り塩漬け..........P172

[Ingredients]

2 chickens

[Marinade]

5 tbsps white peppercorns
5 tbsps sea salt

[Method]

1. Rinse the chicken. Cut off the head and tail. Cut into halves along the breast bone. Wipe dry with towel or paper towel. Set aside.

2. Crush the peppercorns in a mortar and pestle. Stir fry peppercorns and sea salt in a dry wok until fragrant and lightly browned. Remove from heat and leave them to cool. Rub the pepper-salt mixture on both the inside and the outside of the chicken. Put the chicken with the skin down in a strainer with a bottom tray. Then place it on the lower shelf of the fridge (NOT the freezer) for 1 week.

3. The chicken will turn dry and firm after 1 week. Flip the chicken so that the skin faces up. Refrigerate for another week until the chicken in completely dry. Then store the chicken halves in zipper bag separately. Keep them in the freezer.

4. Before serving, take the chicken out of the freezer and let it thaw at room temperature. Then cook according to your preference. Serve.

[Suggested serving method]

a. Rinse the salted chicken briefly to remove excess pepper and salt. Then steam over high heat for 18 to 20 minutes until done. Leave it to cool and chop into pieces. Alternatively, tear the chicken into strips and serve with coriander and shredded spring onion. Drizzle with sesame oil and sprinkle toasted sesames on top. Serve.

b. Make clay-pot rice with the salted chicken.

c. Make salted chicken congee with sliced ginger and spring onion. Feel free to improvise as you like.

[Tips]

1. Before you rub pepper and salt on the chicken, make sure you wipe it dry thoroughly. Any residual moisture on the chicken will extend the drying time and may turn the chicken stale easily.

2. When you stir fry the pepper and salt, make sure you fry them over low heat. This way, the salt would have enough time to pick up the pepper flavour and aroma. The salted chicken will end up tastier.

[材料]

丸鶏...2羽

[下味用漬け汁]

粒コショウ...5大さじ
海塩...5大さじ

[作り方]

1. 丸鶏を洗い、頭、お尻を取り除き、縦半分に切り、チキンペーパー又は布で鶏の水気を拭き取っておく。

2. 粒コショウを押し潰し、海塩と弱火で焦げ始め香ばしくなるまでに炒め、取り出し、冷めてからまんべんなく丸鶏の内外を塗り付け、（鶏の内部は上に向き）通気性のよいザルに盛って、下に受け皿をし、冷蔵庫に入れ一週間放置（冷蔵室に置けてよい、冷凍室に置けないでください）。

3. 一週間後に丸鶏の水分がなくなって硬くなり、丸鶏をひっくり返し、再び冷蔵庫に入れ一週間置き、丸鶏が完全に乾くなって、保鮮用のビニール袋で包み、冷凍室に保存する。

4. 召し上がる時に冷蔵庫から取り出し、室温で自然解凍し、いろんな材料に合せて料理をする。

[おすすめの食べ方]

a. 丸鶏に付いた余分なコショウと塩を洗い取り、水を沸騰してから強火で約18〜20分間に蒸し、冷めてから切り分けて召し上がる。または手で鶏肉を細かく裂いて、パクチ、細切りしたねぎを混ぜ合わせ、上にゴマ油をかけ、香ばしく炒めたゴマを振りかけて召し上がる。

b. 漬けたコショウ塩鶏を土鍋飯に作る。

c. しょうがスライス、段きりにした葱を入れ、塩鶏粥を作れる。いろいろな料理にも合う。

[ポイント]

1. 丸鶏の全身にコショウ、塩を塗り付ける前に、必ず丸鶏に水気を拭き取って、水気が多すぎで、乾かす時間が長く、腐りやすくなる。

2. コショウと塩を炒めるときに、必ず弱火にする。こうすると、塩が十分の時間でコショウの味と香りを吸収し、作った鶏がもっと美味しくなる。

Homemade chopped chilli sauce
自家製の唐辛子ソース..........P180

[Ingredients]

1 kg medium red chillies
50 g old ginger (rinsed, sliced)
80 g garlic (rinsed, sliced)
50 g fermented black beans (rinsed, drain and coarsely chopped)
300 ml oil

[Seasoning]

100 ml white vinegar
40 g coarse salt
5 g chicken bouillon powder (about 1/2 tbsp)
10 g chicken bouillon powder (about 1 tbsp)
90 g chilli oil
60 g sesame oil
100 ml fish sauce

[Method]

1. Rinse the red chillies. Drain well. Cut off the stems and cut them into halves along the length. Seed them. Finely chop them or blend them in a blender.
2. Add white vinegar, coarse salt and 5 g of chicken bouillon powder to the chopped red chillies from step 1. Leave them for 1 day. Drain and squeeze dry. (Discard the liquid.)
3. Heat a wok and add oil. Stir fry sliced ginger over low heat until lightly browned. Add sliced garlic and stir fry until lightly browned. Add fermented black beans and stir fry over low heat until fragrant. Add the chopped chillies. Turn to high heat and stir until the chillies turn soft and the chilli skin begins to roll up. Add chilli oil, sesame oil, fish sauce and 10 g of chicken bouillon powder. Mix well and bring to the boil. Turn off the heat. Leave the resulting mixture to cool.

[Tips]

1. Before using, add 30 g of beer and some sliced garlic to 150 g of chopped chilli sauce. Then spread the resulting mixture over the fish or meat to be steamed. The beer will accentuate the flavour of the sauce and tenderize the meat or fish. It works especially well with beef.

[材料]

赤唐辛子（中）...1000g
ひねショウガ...50g（洗い、スライス切り）
にんにく...80g（洗い、スライス切り）
豆豉(ダウシー)...50g(洗い、水気を取り、粗みじん切り)
油...300ml

[調味料]

白酢...100ml
粗塩...40g
チキンパウダー...5g（約1/2大さじ）
チキンパウダー...10g（約1大さじ）
ラー油...90g
ゴマ油...60g
ナンプラー...100ml

[作り方]

1. 唐辛子を洗い、水気を取り、ヘタを取り、縦半分に切り、ワタと種を取り除き、みじん切り、又はみじん切り器で撹拌する。
2. みじん切りにした唐辛子は白酢、粗塩、チキンパウダー5gを入れ混ぜ、一日漬けた後、出た水分を絞り取る。（水分を捨てる）。
3. フライパンに油を熱し、しょうがを入れ、弱火で香ばしく焼いて黄色になってからにんにくを入れ、また香ばしく焼いて黄色になってから豆豉を入れ、しっかりうま味が出るまで炒める。唐辛子を加え、強火にかけ、唐辛子が柔らかくなり、皮が巻くと、ラー油、ゴマ油、ナンプラー、チキンパウダー10gを入れ混ぜ、沸騰してから火を止め、冷めてから美味しくて万能の自家製唐辛子ソースを出来上がる。

[ポイント]

1. 使用時に毎150g唐辛子ソースにビール30gと適量なにんにくスライスを入れ混ぜ、料理の上にかけて蒸す。ビールを加えたら、食材の鮮度をアップさせ、肉が柔らかくなる。特に牛肉です。

Steamed fish head with homemade chopped chilli sauce
自家製唐辛子ソース入り
コクレンの頭蒸し..........P182

[Ingredients]
1 head of bighead carp (about 900 g)
dried kudzu starch noodles
finely chopped spring onion

[Seasoning]
250 g homemade chopped chilli sauce
50 g beer
2 to 3 cloves garlic (sliced)

[Method]
1. Soak the dried kudzu starch noodles until tender. Drain. Put the noodles into a deep dish.
2. Remove the gills of the fish head. Rinse well and drain. Put the fish head on top of noodle. Mix the seasoning well. Spread evenly over the fish. Boil water in a wok or steamer. Steam over high heat for 12 to 15 minutes. Sprinkle spring onion on top. Serve.

[材料]
コクレンの頭...1コ（約900g）
葛きり...適量
ねぎ（みじん切り）...適量

[調味料]
自家製の唐辛子ソース...約250g
ビール...約50g
にんにく...2〜3粒（スライス切り）

[作り方]
1. 葛きりを軟らかいまで浸り 深い皿に盛っておく。
2. コクレンの頭はえらを取り除き、よく洗い、水気を取っておく。葛きりにのせ、調味料を混ぜてから、まんべんなくコクレンの頭にかけ、お湯が沸騰してから蒸し器に入れ、強火で約12〜15分間に蒸し、取り出し、適量なねぎを振りかけて召し上がる。

Singaporean pickled green chillies
シンガポール風の
青唐辛子ピクルス...P176

[Ingredients]
2 kg medium-sized green chillies

[Marinade]
1 litre white vinegar
800 g sugar
3 tsps salt

[Method]
1. Add sugar to white vinegar. Cook over low heat until sugar dissolves. Turn off the heat. Leave it to cool. Set aside.
2. Rinse the green chillies. Drain well and cut into rings. Place them into a container. Sprinkle salt on top. Mix well. Leave them for 1 day to draw water out of them.
3. Drain any liquid from the green chillies. Squeeze them dry. Put them into a container. Pour in the vinegar syrup from step 1. Mix well. Leave them to soak for 2 days. Serve.

[材料]
青唐辛子（中）...2000g

[下味用漬け汁]
白酢...1リットル
砂糖...800g
塩...3小さじ

[作り方]
1. 白酢に砂糖を入れ、弱火で砂糖が溶けてから火を閉め、冷ましておく。
2. 青唐辛子を洗い、水気を取り、輪きりし容器に入れ、塩を入れ混ぜ、水を出させる一日置く。
3. 第二日に出た水を捨て、青唐辛子を手で絞り水分を切る。容器に入れ、砂糖が溶けた白酢を入れ混ぜ、約二日間に漬けてから召し上がる。

Dried tangerine peel is an expensive herb and condiment. But actually it's not difficult to make your own from scratch. Just pul a string through the fresh tangerine peels and hang them by the window. Or place them in strainers and put them under direct sunlight until they turn stiff and completely dry (usually for 4 to 5 weeks). Then bake them at 60 to 80 ℃ for 4 to 5 hours to ensure no residual moisture is left in them. Leave them to cool and store them in an airtight container. Keep them in a cool dry place away from the sun for 1 year. Then they are ready to use. Tangerines are in season during the winter months when the weather is dry and windy. That would be the best time to make your own dried tangerine peel.

[Tips]

1. Pick large tangerines with thick skin. The dried tangerine peels would turn out more aromatic and have longer shelf life. Thin peels from small tangerines don't have much essential oil and they tend to crack easily when dried.

2. Dried tangerine peels turn mouldy very easily if not handled properly. In case store-bought dried tangerine peels turn moist and soft, just wipe them down with a dry cloth. Then dry them again in the sun and bake them in an oven. Keep them in an airtight container. Dried tangerine peels should be kept in a dry, well-ventilated place away from the sun.

3. Before using dried tangerine peels in cooking, rinse them well and scrape off the white pith with a small knife. The white pith is said to be damp and hot in nature. Use one or two small pieces for cooking each time. Using too much would make the food bitter.

家でも陳皮を作ることはとても簡単だ。みかんの皮を通して窓に掛け、或いはざるにのせ、バルコニーに干し、みかんの皮がカラカラになるまで（約4～5週間）、オーブンに60℃～80℃で4～5時間焼き、水分を完全に取り除く。冷めてから密閉ガラス容器に保存し、一年後に使用できる。冬はみかんの季節で、風が強く乾燥し、陳皮を作るのが一番よい季節だ。

[ポイント]

1. みかんが大きくて皮が厚いのを選んでください。作った陳皮は香りがよく長く保存できる。小さいみかんは皮が薄くて、香りが足りなく、干した後は破れやすくなる。

2. 陳皮の保存が悪いとカビになりやすく、もし買ってきた陳皮は湿気を帯びて柔らかくなり、まずは布でカビを拭き取り、再びに日天干し、オーブンで乾燥させ、密閉ガラス容器に保存し、通気性のよく乾燥と冷暗所に保存したほうがよい。

3. 陳皮を使用する前に水につけて柔らかくする。洗ってから、皮のわたは湿熱が生じるやすいので小さいナイフで内側の白いわたをこそげ取る。毎回1～2枚を使って十分で、多すぎると作った料理が苦い味がする。

Traditional Cantonese pickles
伝統の広東ピクルス..........P190

[Ingredients]

1 white radish
2 small cucumbers
2 carrots
1 to 2 medium-sized red chillies
3 to 4 cloves garlic

[Marinade]

3 tsps salt
1 litre white vinegar
800 g sugar

[Method]

1. Rinse all ingredients. Peel the white radish and dice it. Seed the cucumbers and cut into strips. Slice the carrot. Cut the red chillies into rings. Peel and slice the garlic. Place all ingredients into a container. Add salt and mix well. Leave them for one day until moisture is drawn out of the ingredients.
2. Add sugar to white vinegar. Cook over low heat until sugar dissolves. Turn off the heat. Leave it to cool.
3. Drain the ingredients from step 1. Squeeze them dry.
4. Transfer the squeezed ingredients into a sealable container. Pour in the vinegar syrup from step 2. Mix well. Leave the ingredients in the syrup for 2 days. Serve. Optionally sprinkle some toasted sesames on top before serving for extra nuttiness.

[Tips]

1. This recipe makes the simplest traditional Cantonese pickles. Feel free to steep other ingredients in the vinegar marinade. It works equally well with white cabbage, yam bean and string beans.

[材料]

大根...1本
キュウリ...2本
人参...2本
唐辛子（中）...1〜2コ
にんにく...3〜4粒

[下味用漬け汁]

塩...3小さじ
白酢...1リットル
砂糖...800g

[作り方]

1. 材料を全部洗い、大根の皮を剥いてさいの目に切る。キュウリの種を取り除いて拍子木切りにする。人参を片切り、唐辛子は輪きり、にんにくを片切り、容器に入れ、塩を加えてよく混ぜ、一日置き水を出させる。
2. 白酢に砂糖を入れ、弱火で砂糖を溶けるまで煮、火を止め、冷ましておく。
3. 第二日に材料から出た水を捨て、材料の水気を絞り出す。
4. 最後に材料を容器に入れ、砂糖の溶けた白酢を入れ混ぜ、約二日間に漬けてから出来上がる。召し上がる時に香ばしく炒めたゴマを入れて、もっと美味しくなる。

[ポイント]

1. これはとても簡単で伝統的な広東ピクルス、自分の好きな野菜を加えて一緒に漬けてもよい。キャベツ、葛芋、インゲン豆など、同様に美味しい。

Pickled lotus root in green plum vinegar
レンコンの梅酢漬け..........P192

[Ingredients]

1 kg lotus root
500 ml homemade green plum vinegar
sugar

[Method]

1. Rinse the lotus root. Peel and slice it. Blanch in boiling water for 30 seconds to a minute. Drain and soak them in iced water immediately until cool. Drain again and set aside.
2. Add some sugar to the green plum vinegar. Adjust the amount of sugar according to your preferred sweetness. Transfer the blanched lotus root into a container. Pour in the green plum vinegar. Mix well. Leave it for 1 day and serve.

[材料]

レンコン...約1000g
梅酢...500ml
砂糖...適量

[作り方]

1. レンコンを洗い、皮を剥き、片切り、お湯に10数秒さらしてから、すぐに取り出して氷水に入れて冷まし、水気を取っておく。
2. 梅酢に適量な砂糖を入れ、自分の好みな濃度を調節し、容器にレンコンを入れ、梅酢を入れ混ぜ、約一日に漬けてから召し上がる。

Soy-pickled baby cucumbers
キュウリの醤油漬け..........P194

[Ingredients]

1 kg hothouse baby cucumbers
2 tsps salt

[Marinade]

150 g Maggi's seasoning
100 g rice wine
100 g cold drinking water
100 g sugar
2 slices liquorice

[Method]

1. Rinse the cucumbers and wipe dry. Cut into short lengths. Add salt to cucumbers and mix well. Leave them overnight to draw water out of the cucumbers.
2. Drain the cucumbers the next day. Squeeze them dry. Mix the marinade ingredients well. Put in the cucumbers. Leave them for 2 days. Serve.

[Tips]

1. Do not use tap water in the marinade. It has to be boiled and then cooled to room temperature. Unboiled water carries impurities and micro-organisms. Any unboiled water in the pickles will shorten their shelf life as they turn stale easily.

[材料]

温室キュウリ...1000g
塩...2小さじ

[下味用漬け汁]

美極鮮醤油（マギーソース）...150g
米酒...100g
冷水...100g
砂糖...100g
甘草（カンゾウ）...2枚

[作り方]

1. キュウリを洗い、水気を拭き取り、段切りし、塩を入れ混ぜ、一晩置く水を出させる。
2. 第二日にキュウリを取り出し、手で水分を絞り出し、下味用漬け汁をよく混ぜ、キュウリを入れ、約2日間に漬けてから召し上がる。

[ポイント]

1. 下味用漬け汁の中に水は必ず沸かした水を使用し、生水で使用すると雑質と細菌が含まれ、漬けたキュウリが腐りやすく、カビが生えやすい。

[Ingredients]

1 large Tianjin white cabbage (about 2 kg)

3 to 4 medium-sized red and green chillies (adjust quantity according to your preferred piquancy)

3 cloves garlic

3 tsps salt

[Marinade]

1 litre white vinegar
800 g sugar

[Method]

1. Add sugar to white vinegar. Cook over low heat until sugar dissolves. Turn off the heat. Leave it to cool.

2. Rinse the cabbage and drain well. Cut it into halves. Set aside. Rinse the chillies and garlic. Wipe dry and slice them. Place cabbage, chillies and garlic into a large container. Sprinkle salt evenly in between each cabbage leaf. Rub the cabbage gently until soft. Leave them for 1 day to draw moisture out.

3. Drain all the liquid in the bowl the next day. Squeeze dry the cabbage leaves. Place all ingredients into a large tray. Pour in the pickling vinegar from step 1. Mix well. Leave them for 2 days and serve.

[材料]

白菜...1本(約2000g)
中型の青・赤唐辛子...各3〜4コ(好みの辛さに加減する)
にんにく...3粒
塩...3小さじ

[下味用漬け汁]

白酢...1リットル
砂糖...800g

[作り方]

1. 白酢に砂糖を入れ、弱火で砂糖が完全に溶けるまで煮て、冷ましておく。

2. 白菜を洗い、水気を取り、縦半分に切る。青・赤唐辛子、にんにくを洗い、水気を取り、スライスに切り、容器に入れ、白菜の全身に塩を振りかけ、手で白菜を柔らかくなるまで押し揉み、水分を出させるように一日放置する。

3. 翌日に出た水分を捨て、さらに水分を搾り出し、全部の材料を鍋に入れ、砂糖の溶けた白酢を入れ混ぜ、約2日間漬けてから出来上がる。

[Ingredients]

1 cauliflower (about 700 g)
2 tsps salt
toasted sesames
sesame oil

[Marinade]

1 bottle Chilli sauce for Hainan chicken (230 g)
2 tbsps sugar

[Method]

1. Rinse the cauliflower and cut into florets. Add salt and mix well. Leave it for 1 day until moisture is drawn out. Gently stir and roll the cauliflower from time to time. This way, drier and firmer cauliflower with less water content will still be soft.
2. Blanch the cauliflower in boiling water the next day. Then immediately plunge them into iced water. Leave it until completely cooled. Squeeze dry and transfer into a tray. Add marinade and mix well. Leave it for 1 day.
3. Before serving, sprinkle some toasted sesames and sesame oil on top.

[Tips]

1. The sourness and piquancy of Chilli sauce for Hainan chicken vary from brand to brand. Thus, you may adjust the amount of sugar used according to your preference. Or, you may even make it sourer by adding vinegar or lime juice.
2. Chilli sauce for Hainan chicken is commonly available from stores specializing Southeast Asian grocery which also carry coconut milk, curry and spices. Besides cauliflower, you may also use the same marinade to pickle any gourd or vegetables of your choice.

[材料]

カリフラワー...1コ（約700g）
塩...2小さじ
炒めたゴマ...少々
ごま油...適量

[下味用漬け汁]

海南雞辣椒醬(ハイナン鶏チリソース)...1本
　(230g)
砂糖...2大さじ

[作り方]

1. カリフラワーを洗い、小房に切り分け、塩を入れ混ぜ、一日置き水を出させる。このうちにカリフラワーを軽く数回揉み、こうすると水気をもっと出てしんなりとする。
2. 第二日にカリフラワーを湯切りし、すぐ氷水に浸り、カリフラワーを冷却した後に水気を絞り、容器に入れ、下味用漬け汁を入れ混ぜ、一日に漬けてから召し上がる。
3. 適量の炒めたゴマとごま油少々を入れて召し上がる。

[ポイント]

1. メーカーによって海南雞辣椒醬の辛さとすっぱさが違い、人の好みに砂糖の分量を加減する。又は白酢、ライム汁を加えてもよい。
2. 海南雞辣椒醬は東南アジアのスーパーで買える。このタレを使ってカリフラワーを漬けるだけでなく、自分の好きな野菜を漬けてもよい。

Wine-marinated abalones in distiller's grain sauce
糟滷（酒粕）酔っぱらいあわび..........P200

[Ingredients]

2 Australian abalones (about 600 g each)
5 slices ginger
3 to 4 sprigs spring onion
2 litres chicken stock
200 g rock sugar

[Marinade]

1 bottle Shanghainese distiller's grain marinade
(about 500 ml)
350 ml Shaoxing wine
150 ml chicken stock in which the abalones are
cooked
100 ml cold drinking water
50 ml Sichuan pepper oil

[Method]

1. Shell the abalones. Remove the innards. Scrub
   well. Blanch in boiling water with ginger and
   spring onion.
2. Cook the abalones in chicken stock and
   rock sugar over low heat for 4 hours until soft.
   Leave them to cool. Set aside.
3. Mix the marinade well. Soak the abalones in
   the marinade overnight. Slice the abalones
   and serve.

[Tips]

1. Leave the abalones in the marinade for at
   least 8 hours for them to be flavourful. Yet,
   do not leave them in the marinade for over
   24 hours. Otherwise, they will be too salty. If
   you're not serving the abalones right away,
   take them out of the marinade and store
   them separately for later use.
2. When you stew the abalones, do not add
   any seasoning of salty flavour. Otherwise,
   the abalones will shrink and they take much
   longer to get tender.

[材料]

オーストラリア産の活あわび...2匹(約1200g)
しょうが（スライス切り）...約5枚
ねぎ...3～4本
チキンスープ...1200リットル
氷砂糖...200g

[下味用漬け汁]

上海糟滷（酒粕ベースの汁）...1本（約500ml）
紹興酒...350ml
煮た鮑のチキンスープ...150ml
冷水...100ml
花椒油（ファー　ジョウ　ユ）...50ml

[作り方]

1. あわびの身を殻からはずし、内臓を取り去る。
   たわしでよく洗い、しょうねぎとしょうがの熱
   湯で鮑を霜降りにする。
2. 鍋にチキンスープ、氷砂糖を入れ、弱火で柔
   らかくなるまで約4時間に煮込む。冷ましてお
   く。
3. 下味用つけ汁をよく混ぜ、あわびを入れ、一晩
   に漬ける。第二日に鮑を取り出し、スライスし
   て召し上がる。

[ポイント]

1. あわびに味を染み込ませるように少なくても8
   時間漬ける。しかし、一日以上漬けないように
   してください。漬け過ぎるとしょっぱいにな
   る。すぐに召し上がらなければ、漬け込んだ後
   あわびを取り出して別に保存する。
2. あわびを煮込み時にしょっぱい味の調味料を入
   れないで、そうしなければあわびの身が縮まる
   し、柔らかくしにくい。

[Ingredients]

100 g dried scallops
100 g dried shrimps
50 g bird's eye chillies
300 g shallots
150 g garlic
700 ml oil

[Seasoning]

20 g sugar
120 g satay sauce

[Method]

1. Rinse the dried scallops and dried shrimps. Soak them in water overnight until soft. Break the dried scallops into shreds. Grind up the dried shrimps with mortar and pestle. Rinse the bird's eye chillies and cut into rings. Rinse the shallots. Peel and slice them. Finely chop the garlic. Set aside.

2. Heat oil in a wok. Stir fry bird's eye chillies, shallot and garlic over low heat until fragrant. When they turn dry and golden, put in the dried scallops and dried shrimps. Stir until ingredients are dry and golden. When the mixture starts bubbling, it is almost done. Add satay sauce and sugar. Stir well. Bring to the boil. Turn off the heat. Leave the mixture to cool completely. Store in sealed bottles.

[材料]

干し貝柱...100g
干しエビ...100g
唐辛子...50g
エシャロット...300g
にんにく...150g
油...700ml

[調味料]

砂糖...20g
サテーソース...120g

[作り方]

1. 干し貝柱、干しエビを洗い、水に一晩に柔らかくなるまで浸り、干し貝柱を手でほぐし、石のすり鉢で干しエビを細かく潰し、唐辛子を洗い、細かく輪きり、エシャロットを洗い、皮を剥き、スライス切り、にんにくをみじん切っておく。

2. 鍋に油を熱し、弱火で唐辛子、エシャロット、にんにくを香ばしく炒め、材料の水分が飛ばし黄色になり始めの時、干し貝柱、干しエビを入れて混ぜ炒める。乾くなったり黄色になったりし、泡が立ち始める。この時はそろそろ出来上がる。最後にサテーソース、砂糖を入れて混ぜ炒める。油が再沸騰してから火を止め、盛って冷めてから容器に入れて保存する。自家製サテーXO醬が出来上がる。

Radish pickles with Sichuan peppercorns
大根の麻辣(まーらー)漬け..........P204

[Ingredients]

2.4 kg white radish
2 tsps salt (for marinating the radish in step 1)

[Sichuanese pickling vinegar]

3 cloves garlic (sliced)
5 tbsps Sichuan peppercorns
8 tbsps spicy bean sauce
3 tbsps chilli oil (adjust the amount according to your preferred piquancy)
2 tsps salt
450 g sugar
700 ml white vinegar
4 tbsps sesame oil

[Method]

1. Rinse the radish and wipe dry. Peel and cut into strips about 2 inches long. Place radish in a deep tray. Add 2 tsps of salt and mix well. Leave it for 3 hours to draw the moisture out.
2. Drain any liquid in the tray. Squeeze the radish dry.
3. Mix the Sichuanese pickling vinegar well. Put in the radish and mix well. Leave it for 1 to 2 days. Serve.

[材料]

大根...約2400g
塩...2小さじ(作り方1. 大根の漬け用)

[麻辣(まーらー)下味用つけ汁]

にんにく...3粒(片切り)
花椒...5大さじ
豆板醤...8大さじ
ラー油...3大さじ(お好みの辛さに応じて)
塩...2小さじ
砂糖...450g
白酢...700ml
マー油...4大さじ

[作り方]

1. 大根を洗い、水気を取り、皮を削り、約5cmに拍子木切り、皿に入れ、塩2小さじを大根と入れ混ぜ、約3時間漬ける。
2. 3時間後、大根から出た水分を切り、手で大根の水分を搾り出す。
3. 麻辣(まーらー)下味用つけ汁を混ぜ合わせ、大根を入れ混ぜ、1〜2日間漬けてから召し上がる。

Miso-salted egg yolk
アヒル卵黄の味噌漬け..........P206

[Ingredients]

9 duck eggs

[Marinade]

1 pack low-salt miso (Japanese fermented soybean paste, about 600 g)
2 tbsps rice wine
5 tbsps sugar

[Method]

1. Refrigerate the duck eggs for half a day before use. This way, the egg yolks will be slightly set and are easier to handle.
2. Mix the marinade well. Spread a layer of marinade on the bottom of a deep container. Then put half of the cheesecloth on top. With the rounder end of a duck egg, make 9 round notches on the marinade. Each notch should be enough to hold an egg yolk. Then crack the duck eggs. Discard the whites and arrange one yolk into each notch on the marinade. Fold in the remaining half of the cheesecloth to cover the yolks. Then carefully spread another layer of marinade on top of the cheesecloth. Seal the container with cling film. Keep in the fridge. Try not to move it as far as possible. Leave them for 3 weeks or so, by which time the egg yolks should be set. They are then ready to be served.

[Tips]

1. Miso-salted egg yolks can be steamed and served with congee or savoury soybean milk. Or, you can use it to replace salted duck egg yolks in various recipes, such as deep-fried prawns, crabs, bitter melon or bamboo shoots in salted egg yolk sauce, etc.
2. The top layer of marinade over the cheesecloth should only be 5mm in thickness. If it's too thick, too much pressure will be put on the egg yolks which may burst as a result.
3. Before you put the egg yolks into the notches, try to remove the egg white as much as you can. Too much egg white will affect the salting process.
4. There are different kinds of miso in the market. You may pick any kind according to your own taste. Different miso actually makes salted egg yolks of different flavours. However, try to use low-salt miso for the best results. It's because the miso comes in direct contact with the egg yolks without being separated by the shell. Egg yolks salted this way in traditional miso might turn out too salty.

[材料]

アヒル卵...9コ

[下味用漬け汁]

日本産の低塩味噌...1コ（約600g）
米酒...2大さじ
砂糖...5大さじ

[作り方]

1. 漬ける前に、漬けをしやすくするため、アヒル卵を冷蔵庫に半日置き、卵黄を少し固めにする。
2. 下味用漬け汁をよく混ぜ合わせ、深い皿に下味用漬け汁を平に敷き詰め、ガーゼを敷き、アヒル卵の下部分（丸いほう）で、ガーゼの上から軽く押し、くぼみをつけ、アヒル卵を割って、卵黄だけを取り出し、卵黄をくぼみに落とし入れガーゼをかぶせる。上にまた下味用漬け汁をまんべんなく敷き詰め、ラップをし、冷蔵庫に入れる。そのまま3週間ほど置き、卵黄が固まって固体になると、料理に使える。

[ポイント]

1. 漬け上がった卵黄を蒸して、お粥や塩味の豆乳などに合わせられる。そして、伝統的な塩漬け卵黄を代わっていろんな料理に使えられる。卵黄エビ、卵黄カニ、卵黄ゴーヤ、卵黄冬筍（卵黄と冬竹の子）など………
2. 最後にガーゼに下味用漬け汁を塗る時に、厚くない方がよい（約5mmは1分）、厚く過ぎると重さが卵黄を潰しやすくなる。
3. 卵黄をくぼみに入れる前に、卵白をできるだけ全部取り去り、湿気が多いと漬けた卵黄の効果が落ちる。
4. 市販にいろいろな日本味噌があり、風味が違い、自分の好みで選び、味噌の味が違うと、漬けた卵黄の味も違う。直接に卵黄を漬けるから伝統的な味噌で漬けるとしょっぱい過ぎるかもしれない。低塩の味噌を選んだほうがよい。

Miso-salted egg yolk crusted soft-shell crabs
味噌卵黄のソフトシェルクラブ..........P208

[Ingredients]
4 soft-shell crabs
8 miso-salted egg yolks
caltrop starch (for coating the crabs)

[Deep frying batter]
150 g flour
1 g baking powder
10 g caltrop starch
water

[Method]
1. Thaw the soft-shell crabs and remove the gills. Rinse and squeeze dry. Cut each crab into halves. Set aside. Steam the miso-salted egg yolks for 10 minutes until done. Mash coarsely with a fork.
2. Mix the deep frying batter ingredients together. Add water slowly until it's of a consistency that coats your finger well and when you lift the batter with your finger it drips like a thread. Then put in the mashed miso-salted egg yolks from step 1. Mix well.
3. Coat the crabs lightly with caltrop starch. Then dip them into the deep frying batter from step 2. Heat oil in a wok up to 180℃. Deep fry until golden. Serve.

[Tips]
1. When you make the deep frying batter, make sure it's not too runny. Otherwise, the miso-salted egg yolks cannot adhere to the crispy crust, ending up being deep-fried on their own in the oil. Besides, the miso-salted egg yolks should not be mashed too finely either. I prefer them in small lumps which accentuate the mouthfeel.
2. There are some tricks to crispy deep-fried goods. Most chefs in a professional kitchen will deep fry the food over medium heat until the food is golden. Then they turn the heat up and fry the food over high heat for 10 seconds or so. The high temperature in the last stage kind of pushes the moisture out of the crust, so that the fried food would taste less greasy and the crust would be fluffier. For similar results in a household kitchen, we usually deep fry the food once and leave it slightly undercooked. Set the food aside for a few minutes. Then we deep fry it in oil once more.
3. Some major Japanese supermarkets carry deep frying batter mixes in a box. They actually work quite well. If you don't want to make it from scratch, you can get them pre-mixed. Just add water and dunk your food in.

[材料]
ソフトシェルクラブ...4匹
味噌漬け卵黄...8コ
片栗粉（クラブを揚げる時に使用）...適量

[揚げ物の衣]
小麦粉...150g
ベーキングパウダー...1g
片栗粉...10g
水...適量

[作り方]
1. ソフトシェルクラブを解凍し、えらを取り除き、よく洗って水気を取り、胴体を半分に切り離す。味噌漬け卵黄を約10分間に蒸し、取り出してからフォークで押し潰しておく。
2. 衣をよく混ぜ、ゆっくり水を加え、衣の濃度は指で持ち上げると、滑らかに流れ落ちる程度に、押し潰した卵黄を入れ混ぜて衣になる
3. ソフトシェルクラブに適量な片栗粉をかけ、衣をたっぷりつけ、180度に温めた油でかりっとするまで揚げてから召し上がる。

[ポイント]
1. 衣は必ず濃くなる状態にし、薄すぎると、卵黄が衣をうまくつけられない、揚げてる途中で分解しやすくなる。その上、卵黄を潰しすぎると食感が分からなくなる。
2. かりっとした衣の作り技：一般のレストランは料理人が中火で揚げ、黄色になり始める時、強火にかけ、油の温度が上がると、あど10数秒ぐらい揚げる。高温の油が揚げ物の中にある水分を蒸発し、こうしたら、揚げたものはかりっとなる。口の中に油っぽくならない。一般の家庭では揚げ物を作る場合、もうすぐ揚げ上がりのものを取り出し、数分間を置いてから、もう一回油を熱して揚げる。こうすると、レストランと同じくかりっと揚げ効果がある。
3. 日本系の大型スーパーに売っている揚げ粉は、揚げた効果もよい。もし便利にしたいなら、それを使ってもよい。

Grilled cod in miso sauce
タラの味噌漬け..........P210

[Ingredients]

2 pieces cod fillet

[Marinade (A)]

2 tbsps sake (Japanese rice wine)

[Marinade (B)]

900 g miso (Japanese fermented bean sauce)
250 g sugar
150 g mirin (Japanese sweet cooking wine)

[Method]

1. Scale the cod and rinse well. Wipe with paper towel. Brush sake evenly on both sides of the cod. Set aside.
2. Mix the marinade (B) well. Spread a layer of marinade (B) on the bottom of a dish dish. Place half of a piece of cheesecloth over it. Put the fish over the cheesecloth. Then put the remaining half of the cheesecloth on top. Top with another layer of marinade (B) on top. Leave them for a night.
3. Lift the cheesecloth and remove the fish from the marinade. Put the cod fillet on a grill pan. Bake in an oven at 250℃ for 6 to 7 minutes until the cod is lightly browned and crispy on the outside. Serve. Optionally, serve with creamy salad dressing and squeeze lime juice over the fish.

[Tips]

1. When you spread a layer of marinade (B) on a dish, make sure the layer is even and smooth. Otherwise the cod cannot be evenly marinated.
2. Do not marinate the cod for longer than 24 hours. Otherwise, it will be too salty. If you don't plan to cook it right away, at least take the cod out of the marinade. This will stop the marinating process so that the fish will not be too salty.

[材料]

タラのステーキ...2切れ

[下味用漬け汁 (A)]

日本清酒...約2大さじ

[下味用漬け汁 (B)]

味噌...900g
砂糖...250g
みりん...150g

[作り方]

1. タラは鱗を取って洗い、キチンペーパーで水気を拭き取り、タラの両面に清酒を塗っておく。
2. 下味用漬け汁（B）を混ぜ合わせ、深い皿に下味用漬け汁を平に敷き詰め、ガーゼを敷き、タラをのせる。また、ガーゼを折ってタラを包み、上に味噌を塗り、一晩に漬ける。
3. 第二日にタラを耐熱皿に入れ、オーブンは250度に予熱し、タラを入れ約6～7分間焼き、きつね色になってから出来上がる。召し上がる時にサラダを合わせ、ライム汁をかけて召し上がる。

[ポイント]

1. 皿に下味用漬け汁は必ず均一な厚さになるように敷き詰める。表面がでこぼこになると、味が不均一になる。
2. タラは漬けすぎ注意。一日以上漬けるとしょっぱいすぎになり、すぐ料理に使わなければ、漬け上がった後に味噌を取り除き、タラを別に保存する。

享受孤獨愛漫 • 遊

我的第一次是在十七歲！

是第一次自己一個人去旅行……噢！已經是那麼遙遠的事了，還記得那是第一年出來做事，買了第一部Nikon FM相機，只是拿了一個背包、一部相機，從香港坐火車去深圳，深圳坐火車去廣州，再從廣州坐了二十二小時的火車去桂林，一個人自由自在玩了九天。那次是我人生中一次很難忘的經驗之一。

直到現在，我也會在適當的時候，找一個喜歡的地方，安排給自己一個獨處的假期，平平靜靜的渡過幾天。大家不要誤會，我不是那種不喜歡合群兼且性格孤僻的人，在我身邊有很多很好的朋友，我亦很享受跟心愛的人或是三五知己一起去旅行，同時，我亦有好幾位非常要好，去旅行的「腳」！但是，偶爾一個人去「流浪」的感覺，對我而言，實在有說不出的美妙與快感！因為從我出來做事至今，從事廣告及飲食，都是要面對很多、很多不同種類的人，同時我亦有很多出外工作的機會，但每次都是一大隊人，有時不但過於熱鬧，還要應付這樣、處理那樣，久而久之，你會覺得麻木，很想體驗一下自在寧靜的感覺。

每次當我獨自身處異地的時候，我就有一種像出家人的感覺，了無牽掛，心無雜念，喜歡去那裏便去那裏，不需要像平常工作般要有詳細的計劃及安排，沒有時間性的壓力，除了坐車、問路或點餐之外，我可以整天不用跟別人講說話，不用虛偽。我還發現，獨處時刻的感受還會更加敏銳，讓你看到的事和物感覺都更加細膩及清晰。去年，我便抽了六天，一個人專程去了台北，跟咖啡達人林東源學咖啡沖調及拉花，那種像「遊學」的感覺，非常寫意及難忘。

過去了的新春假期，我便一個人跑了去台北，訂了間好的酒店，然後把電話關掉，每天一早起床，到酒店餐廳吃一頓豐富早餐，然後上上網，看看新聞及Facebook，接着到誠品書店逛逛，打書釘，買文具，那裏有很多各式各樣不同種類的食譜，在那裏，已經可令我呆足一天，我常跟朋友說，台北信義區的誠品書店，簡直是我的天堂！我還去了淡水，看街頭藝人賣藝獻唱，吃一些自己喜歡的地道小食。我更一個人坐高鐵走到台中「遊」了一天，沒有預先設定的行程，看到想吃的小攤子便坐下來嘗試一下，看到喜歡的咖啡室又可以在那裏休閒地坐上兩個鐘，看着旁邊走過不同的路人，感受一下當地的生活氣息……會心微笑。獨處外地，還有一件我很喜歡做的事，就是安排自己去一些當地特別、地道而又很想試的餐廳毫不節制地大吃一頓，感覺實在又幸福！

若你從未有過這種體驗的話，我認為確是一種遺憾！給自己一個機會，像我一樣，感受做一次浪人的樂趣！

惜飲惜食

曾經聽過這樣一個故事：有一個人，他母親去世後仍然十分想念着他的母親，很想知道母親在泉下的情況，於是他找來靈媒，為他安排了陰間遊，去到陰間，見到了一個讓他很震驚情景，他看到母親正很痛苦地吃着一大桶比人還要高的東西，他問母親：媽，你在吃甚麼？母親說：我在吃生前我所吃剩及曾經浪費了的食物；然後兒子看見母親旁邊還放着另一大桶腐爛食物，他再問母親：這桶又是甚麼呢？母親答：仔呀！這些就是你到目前為止所吃剩的食物。

這聽起來是一個警世寓言故事，但當中含意也相當發人深省！我出生在佛山一個小康家庭，正值是那個火紅紅的文化大革命時代，在那個時候，食物物資相當缺乏，很多基本食品都要配給，買米要糧票、買肉要肉票、買魚要魚票…購買其他糧油雜貨都要一些指定的「票」，還要根據每個家庭人口每月每人限額分配，購買時還要排上好幾小時的隊才能買到。還好，我家有一個「港澳同胞」身份的爸爸，可以定期將一些「補給物資」寄回家中，在那時候很偶爾能吃上一些牛奶糖、威化餅之類的零食已經足以開心一整天。一個蘋果很多時都要三兄弟妹共分，我還很清楚記得小時候在「托兒所」每星期才有一次有肉的湯，而每碗湯裏面的那塊肉湯渣，都是小朋友們期待的美食，因此「食物」對我們這些經歷過艱難歲月的人來說是非常難得及珍惜的。

而諷刺的，是那麼多年後的今天，我的工作亦與食物結下了不解之緣。亦因為如此，這些與食物有關的工作，都有機會浪費了很多食物，例如拍攝飲食節目、出席烹飪比賽、製作食譜、試菜等等，雖然我都在可能的情況下把食物食完，將可以保留的食物保存或分給其他人，但有時這些工作範疇亦非我個人控制之內。因此，每每遇到這種情況都令我非常無奈及痛心，我真擔心會有一天，會落得故事主角般的下場！

然而，現今社會豐衣足食，尤其生活在香港，大部分人都毋須為食物而操心。年輕一代很多都沒有經歷過像我們童年時候的日子，食物是理所當然的事。加上近年飲食文化普及，很多香港人在外出用膳、應酬、飲宴等場合，都不停浪費了很多食物，而這些本來可以在很多貧窮地方可以養活很多人的食物，最終都是變成了食物垃圾，被運到堆填區去。

其實很多人都已經知道，在世界上有很多地方、很多人都生活在貧窮線以下，長期都生活在飢餓的日子之中，若果這些被浪費掉的食物沒被浪費，或許可以挽回很多生命。很希望大家在享受美食的同時，對食物應有一種珍惜的飲食態度，讓食物能發揮其最應有的功能，令更多、更多、更多的人可以分享！

隨時使用手機，
掃描食譜內的 QR 碼 (Quick Response Code)，
連接上網，即時收看 Jacky 的烹飪步驟短片，
隨時隨地輕鬆入廚！

QR 碼 上 網
互 動 資 訊

睇片教用 QR 碼：
http://www.facebook.com/formspub
http://www.formspub.com/

Enjoy Cooking. Enjoy Life.

vitavit ® *comfort*
pressure cooker

vitavit ® *premium*
pressure cooker

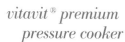

WWF *for a living planet*

保育海洋生態
作出明智選擇

Conserve our marine life

Make a smart choice

海鮮
選擇指引
Seafood
Guide

食用可持續生產的海鮮！
Go for sustainable seafood!

電話查詢 Enquiry: (852) 2526 1011

下載海鮮選擇指引 Download the Guide   **w w f . o r g . h k / s e a f o o d**

TERRA.
chips

非凡享受

Naturally Tasty · Never Ordinary

地中海異國風情蔬菜片 　　　　　　藍薯片

天然 · 健康 · 精明選
Natural · Healthy · Your Choice

Imagine 有機走地雞湯　　　　　　合成味精雞湯

真「證」有機原味，你試過未？

Imagine 有機走地雞湯，以精選有機健康走地靚雞、有機蔬菜及香料、優質海鹽等7種天然食材，並以經過濾的食水熬成。走地雞味鮮且香，健康低脂，啖啖靚雞原味，真材實料，無須添加味精與人造成份，鮮雞美味截然不同，其他雞湯無法比擬。

真「證」有機

Imagine 有機走地雞湯，從嚴選原材料以至整個製作過程，均符合嚴謹有機準則，並獲頒美國農業部(USDA)有機認證，證明真正有機，更可信賴，更加健康。

美國農業部有機認證

健康創新名廚
Jacky Yu 推介

天然煮意鮮湯系列

有機素雞湯　　有機走地雞湯(低鈉)　　有機蕃茄濃湯　　有機西蘭花濃湯　　有機粟米濃湯(輕鬆)

有機走地雞湯

天然·健康·精明選
Natural · Healthy · Your Choice

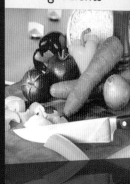

SPCA
愛護動物協會

Say 'No' to sharkfin soup
請向魚翅說「不」

我們為什麼不應吃魚翅？

- 取翅是一個極其殘忍的過程，要在牠們還活時砍取魚鰭。被砍傷的鯊魚會被擲回海裡，掙扎致死。

- 每年有1億條鯊魚被殺，而7千3百萬條是為了製造魚翅。

我們如何可以拯救鯊魚？

- 不食用魚翅及其他鯊魚製品。

- 轉告親友有關鯊魚的困境，並向他們游說不再食用魚翅。

- 支持香港及國際機構倡議禁止魚翅買賣。

如需更多信息，請瀏覽www.spca.org.hk/sharksfin

For more information, please visit www.spca.org.hk/sharksfin

魚翅以外的佳餚精選

雖然中國人慣於在佳節奉上魚翅湯羹，但亦絕對能以善待動物的菜式所取代。囍宴創辦人Jacky亦絕對認同，並推介材料包括瑤柱、蟹肉、蝦、帶子、海魚、魷魚、冬菇及蠔豉的瑤柱海鮮羹。Jacky相信煮出美味佳餚的同時，亦絕對能兼顧尊重生命、愛惜生態環境的訊息。

囍宴

xǐ yàn

xǐ yàn

囍宴 廚 • 藝 (私房菜)
香港 灣仔 灣仔道83號 3樓

XǏ Yàn
3/F, 83 Wanchai Road, Wanchai, Hong Kong
Tel: 2575-6966 Fax: 2575-6928
Email: resv@xiyan.com.hk

囍宴 甜 • 艺
香港 灣仔 永豐街8號 地下1號舖

XǏ Yàn • Sweets
Shop 1, G/F, 8 Wing Fung Street,
Wanchai, Hong Kong
Tel: 2833-6299 Fax: 2833-6696
Email: sweets@xiyan.com.hk

囍宴 東 • 丰
香港 太古城第五期 寧安閣G505 & 508 號舖

XǏ Yàn • East
Shop G505 & 508, Ning On Mansion
Stage V Taikoo Shing, Hong Kong
Tel: 2380-0919 Fax: 2380-0396
Email: east@xiyan.com.hk

囍宴 廚 • 坊
香港黃竹坑黃竹坑道59-61號
本利發工業大廈6樓

XǏ Yàn Penthouse
6/F., Benefit Industrial Factory Building,
59-61 Wong Chuk Hang Road,
Wong Chuk Hang, Hong Kong
Tel: 3622-3912 Fax: 3622-3910
Email: penthouse@xiyan.com.hk

XǏ Yàn • Singapore
38A Craig Road, Singapore 089676
Tel: +65 6220-3546 Fax: +65 6220-7069
Email: info@xiyan.com.sg

XǏ Yàn • Shaw
1 Scotts Road, Shaw Centre, #03-12/13,
Singapore 228208
Tel: +65 6733 3476
Email: shaw@xiyan.com.sg

囍艺 • 上海
上海市徐匯區淮海中路999號
環貿iapm商場L5-510

Xi Yi
L5-510, iapm mall, No. 999 Huai Hai Road
Middle, Xu Hui District, Shanghai, PRC
Tel: +021 6203-6138
Fax: +021 5450-0120
Email: xiyi@sh-xiyan.com

xǐ yàn 38A

極品XO醬 (150g)
XO sauce

口水雞麻辣油 (140g)
hot & spicy fragrant oil

養生私房老火靚湯
Xi Yan homemade soup

黃金一口鮑 (6-8隻/pcs)
golden abalone in soy sauce

黃金禮盒
Bravo treat

黃金一口鮑(1罐)+鮑魚麵(4個)+蝦子麵(4個)
abalone(1 can)+abalone noodle(4 pcs)+shrimp's roe noodle(4 pcs)

撈囍禮盒
Deluxe treat

極品XO醬(1盒)+鮑魚麵(4個)+蝦子麵(4個)
XO sauce(1 box)+abalone noodle(4 pcs)+shrimp's roe noodle(4 pcs)

擔擔拉麵
Dan Dan noodles

囍宴圍裙(全身)
Xi Yan bib apron

囍宴圍裙(半身)
Xi Yan bistro apron

私房賀年糕點系列
Xi Yan new year cake

金牌蘿蔔糕/有營蘿蔔糕/臘味芋頭糕
桂花年糕/沖繩黑糖年糕/黃金馬蹄糕
supreme radish cake / farmhouse radish cake /
preserved meat taro cake / osmanthus sticky rice cake /
Okinawa black sugar sticky rice cake / water chestnut cake

囍宴購物袋
Xi Yan tote bag

囍宴杯墊(5個)
Xi Yan coaster(5 pcs)

囍宴

xǐ yàn

私房優質產品

現正囍氣熱賣

eShop QR Code

囍宴產品禮券可於網上購買，詳情請瀏覽以下網址
http://www.xiyan.com.hk/eshop

| 作者/ 食物造型設計 | Author/ Food Styling |
| 余健志 | Jacky Yu |

策劃/編輯　Project Editor
Catherine Tam

攝影　Photography
Johnny Han
Jacky Yu

美術統籌及設計　Art Direction & Design
Amelia Loh

出版者　Publisher

Forms Kitchen Publishing Co.,
an imprint of Forms Publications (HK) Co. Ltd.

香港英皇道499號北角工業大廈18樓　18/F, North Point Industrial Building, 499 King's Road, Hong Kong

電話　Tel: 2138 7998

傳真　Fax: 2597 4003

網址　Web Site: http://www.formspub.com

http://www.facebook.com/formspub

電郵　Email: marketing@formspub.com

發行者　Distributor

香港聯合書刊物流有限公司　SUP Publishing Logistics (HK) Ltd.

香港新界大埔汀麗路36號　3/F., C&C Building, 36 Ting Lai Road,

中華商務印刷大廈3字樓　Tai Po, N.T., Hong Kong

電話　Tel: 2150 2100

傳真　Fax: 2407 3062

電郵　Email: info@suplogistics.com.hk

承印者　Printer

合群(中國)印刷包裝有限公司　Powerful (China) Printing & Packing Co., Ltd.

出版日期　Publishing Date

二〇一一年七月第一次印刷　First print in July 2011

二〇一五年七月第二次印刷　Second print in July 2015